Effects of Abatement of Domestic Sewage Pollution in Biscayne Bay

STUDIES IN TROPICAL OCEANOGRAPHY

No. 1. Systematics and Life History of the Great Barracuda, *Sphyraena barracuda* (Walbaum)
By Donald P. de Sylva

No. 2. Distribution and Relative Abundance of Billfishes *(Istiophoridae)* of the Pacific Ocean
By John K. Howard and Shoji Ueyanagi

No. 3. Index to the Genera, Subgenera, and Sections of the Pyrrhophyta
By Alfred R. Loeblich, Jr. and Alfred R. Loeblich, III

No. 4. The R/V *Pillsbury* Deep-Sea Biological Expedition to the Gulf of Guinea, 1964-1965 (Part 1)

No. 5. Proceedings of the International Conference on Tropical Oceanography, November 17-24, 1965, Miami Beach, Florida

No. 6. American Opisthobranch Mollusks
By Eveline Marcus and Ernst Marcus

No. 7. The Systematics of Sympatric Species in West Indian Spatangoids: A Revision of the Genera *Brissopsis, Plethotaenia, Paleopneustes,* and *Saviniaster*
By Richard H. Chesher

No. 8. Stomatopod Crustacea of the Western Atlantic
By Raymond B. Manning

Studies in Tropical Oceanography No. 9
Institute of Marine and Atmospheric Sciences
University of Miami

Effects of Abatement of Domestic Sewage Pollution on the Benthos, Volumes of Zooplankton, and the Fouling Organisms of Biscayne Bay, Florida

By J. Kneeland McNulty

UNIVERSITY OF MIAMI PRESS
Coral Gables, Florida

This volume may be referred to as
Stud. trop. Oceanogr. Miami 9:
107 pp., 19 figs., March, 1970

Library of Congress Catalog Card Number 69-19867

SBN 87024-113-3

Manufactured in the United States of America

Contents

Figures

Tables

Foreword

The menace of pollution is belatedly coming to be appreciated and the urgent need for its study realized. Historically such studies have progressed from fresh waters to estuaries and then to the sea. An understanding of the effects of pollution calls for a comparison of the ecology of a region with and without the pollution. Unfortunately, especially with sewage pollution, the onset is usually gradual, and, by the time that the need for a study is recognized, it is too late to obtain data on the earlier, unpolluted conditions. Biscayne Bay, therefore, afforded the welcome opportunity of studying the conditions in a heavily sewage-polluted estuary and comparing these with what prevailed some years later after a sewage treatment plant was in operation and the pollution in the Bay considerably reduced.

The study was important also because the fauna and flora are tropical, and very little is known of the effects of pollution in the tropics. It is known, however, that there is a tendency for tropical organisms to be more sensitive to departure from optimal conditions than are temperate organisms. This has the advantage of making pollution effects more apparent. On the other hand, there are far more species in the tropics, there have been fewer studies there than in temperate waters, and correspondingly much less is known of the basic biology and ecology of individual tropical species. Again, it is fortunate that many of such studies as have been made have been carried out in Biscayne Bay.

The Department of Health, Education and Welfare appreciated the need for the studies presented here and supported them as well as their publication. The publication of rather extensive original data is justified by the probability that they will be needed for comparative purposes in the future when such observation cannot be repeated because of changed conditions. This situation has, in fact, already arisen with the threat of thermal pollution farther south in the same bay. Dr. McNulty's results should be valuable to both biologists and engineers.

The Editors

Preface

The comparison of ecological conditions in Biscayne Bay, Florida, before and after pollution abatement was conceived initially by Dr. F. G. Walton Smith, Director of the Institute of Marine Sciences, University of Miami, early in the 1950's when the City of Miami announced plans for a large domestic sewage disposal plant. The plant began operations in late 1956. Studies were made before and after pollution abatement to document selected biological and environmental changes. This account is primarily a comparison of the conditions of the benthos, volumes of zooplankton, fouling organisms, and phosphate-phosphorus before and after abatement of pollution, together with a summary of the literature on pollution and the marine environment. A few key freshwater studies are mentioned because they preceded and provided a frame of reference for marine studies.

Acknowledgments

This study, Contribution No. 994 from the Institute of Marine and Atmospheric Sciences, University of Miami, was supported by a graduate assistantship of the University of Miami and Grants E-510 and RG-4062 from the National Institutes of Health, U.S. Public Health Service. It is based on a dissertation submitted in partial fulfillment of the requirements for the degree of Doctor of Philosophy at the University of Miami, Coral Gables, Florida.

It is a pleasure to acknowledge the assistance and cooperation of many people at the Institute of Marine and Atmospheric Sciences, University of Miami. I am grateful for the continued interest of Dr. F. G. Walton Smith, Director of the Institute. Special appreciation is extended to Dr. Hilary B. Moore, chairman of my Ph.D. graduate committee, for his help over the years. I am indebted to the other graduate committee members for their constant advice and encouragement. They are Drs. Eugene E. Corcoran, Leonard J. Greenfield, Sheldon B. Greer, Clarence P. Idyll, Harding B. Owre, and Gilbert L. Voss. Grateful appreciation is extended to Mrs. Maria Foyo for sorting and preliminary identification of the benthos; to Mr. N. Kenneth Ebbs, Jr., Dr. Raymond B. Manning, Dr. Lowell P. Thomas, Dr. J. W. Wacasey, and to Mr. Robert C. Work for help with final identification; to Mr. John G. Stimpson for the dissolved phosphate-phosphorus determinations; to Mr. Robert W. Thornhill for help with construction of material; to Mrs. Mary G. Romine, Mrs. Elizabeth W. Moore, and Rita B. McNulty for their typing; and to Mr. George R. Coslow of the United States Army Corps of Engineers for data on the history of dredging in Biscayne Bay. Finally, I am indebted most of all to my wife, Lois, who has assisted in every possible way.

Effects of Abatement of Domestic Sewage Pollution in Biscayne Bay

Introduction

Pollution and the Marine Environment

Athelstan Spilhaus, chairman of the committee on pollution of the National Academy of Sciences, has said that pollution is the inevitable consequence of an overconcentration of people. It occurs when the waste products from certain human activities alter the environment to the detriment of other human activities. Alterations imposed on the biota are the special concern of the biologist. At least from the biologist's point of view, pollution occurs when the wastes from human activities have a detrimental effect on the biota.

Effects of pollution depend upon the kinds and amounts of pollutants and the physical and biological characteristics of the receiving waters. Kolkwitz & Marsson (1908, 1909) are credited with the first studies of the effects of pollution on aquatic plants and animals (Gaufin, 1957). They divided the biota into three groups: those adaptable to only slight pollution–in this case organic enrichment–which they called "oligosaprobic" organisms; those adaptable to medium pollution–the "mesosaprobic" organisms; and those adaptable to a high degree of pollution–the "polysaprobic" group of organisms.

A more complex system of classification was proposed by Richardson (1921, 1928) as a result of studies on the shore and bottom fauna of the Illinois River and its connecting lakes. He discerned seven groups of macroinvertebrates: I, the pollution tolerant species; II, III, and IV, subpollution groups containing unusually tolerant, unusually tolerant or doubtful, and less tolerant species, respectively; V, pulmonate snails and air-breathing insects; VI, current-loving species other than pulmonate snails and air-breathing insects; and VII, cleaner water species.

At the time of Richardson's studies extermination of many species had already occurred in polluted parts of the Illinois River (Baker, 1926). Despite growing scientific and public awareness of the dangers of pollution, it remained for post-World War II biologists to document thoroughly the course of destruction. In 1948, the Academy of Natural Sciences of Philadelphia undertook a comprehensive study of the Conestoga Creek, Lancaster County, Pennsylvania, the results of which provided the basis for a biological measure of stream pollution proposed by Ruth Patrick (1949). In her method, the numbers of species in various groupings, some pollution tolerant and some intolerant, were displayed in histograms. Histograms for various stations in the pollution gradient showed characteristic patterns. Wurtz (1955) made some improvements on the method.

During the same period, biologists at the Taft Center of Sanitary Engineering in Cincinnati, Ohio, were conducting intensive studies in fresh water which culminated in many publications (Bartsch, 1948; Bartsch & Churchill, 1949; Gaufin & Tarzwell, 1952, 1956; Tarzwell & Gaufin, 1953; and many others). It is of special interest that Gaufin & Tarzwell (1952, 1956) recognized indicator communities of macroinvertebrates in addition to the traditional indicator species.

The brief review above represents a very small sampling of the vast literature on biological effects of pollution in fresh water. Two excellent reviews are available (Hynes, 1960; Hawkes, 1962), as well as the thoroughly documented review of the effects of specific pollutants published by the state of California (California State Water Quality Control Board, 1963). The latter review contains a bibliography of 3,827 entries.

Turning to pollution of the sea and coastal areas, it is convenient to divide the marine environment into four categories, after the manner suggested by Mohr (1960). They are: first, offshore areas well away from land; second, offshore areas near land; third, certain harbors and enclosed bays that have moderate to negligible changes in salinity and limited water circulation; and fourth, positive estuaries.

The first category, offshore areas well away from land, includes disposal sites for industrial acids, some radioactive wastes, domestic sewage solids, and other residues. Such acids are deposited from barges from New York, San Francisco, and other large ports. Ketchum (1952) described the dispersion rate of acid-iron wastes from a barge at sea. Certain low-level radioactive wastes packed in suitably strong, durable, weighted containers are at times deposited on the sea floor, care being taken to avoid areas which might be fished by trawlers (F. Koczy, personal communication). Aquatic disposal of radioactive wastes is treated in a number of publications (International Atomic Energy Agency, 1960; National Research Council, 1957, 1959; Saddington & Templeton, 1958; Ingram & Wastler, 1961; Mauchline & Templeton, 1964).

Second, many open coastal areas receive effluents from offshore sewer outfalls and from polluted rivers. The Miami outfall off Virginia Key, for example, is 1,370 meters in length and discharges in 6 meters of water. The Miami Beach outfall extends 2,140 meters seaward and discharges in 12 meters of water (California State Water Pollution Control Board, 1956). Design of the Hyperion outfall off Los Angeles, California, called for one 11 kilometers in length discharging to a minimum depth of 92 meters, while another outfall nearby was designed to extend 8 kilometers seaward to a minimum depth of 61 meters (Hyperion Engineers, 1957). Blegvad (1932) was the first to study biological effects of domestic sewage issuing from an offshore outfall (in the Sound off Copenhagen). Petersen grab samples taken at varying distances from the outfall showed adverse effects on bottom fauna within a radius of 100 to 200 meters. Immediately beyond the adversely affected area he found a comparatively rich fauna dominated by bivalves, especially *Cardium edule, Mya arenaria,* and *Macoma baltica,* often occurring with *Zostera.* An unusually large number of species and unusually high population densities characterized this zone, beyond which normal conditions prevailed.

Because two-thirds of California's municipal and industrial wastes are discharged to saline waters, the state and federal governments have supported jointly a comprehensive program of ecological studies in coastal waters (California State Water Pollution Control Board, 1960). Some studies were concentrated intensively on basically one resource, while others were more extensive. An example of the first type was a study of kelp, in which causes of damage to the kelp were sought. While increased turbidity from pollution may have been a factor, it was found that grazing by sea urchins was a very important factor in the losses (California State Water Pollution Control Board, 1964b). Other studies dealt in a comprehensive way with the interrelationships of pollution and the marine environment. Phytoplankton production was part of one recent study. Near the point of discharge of an offshore outfall there was low production. Downstream from the outfall after a few hours' flow the production was well above normal, following which there was a gradual return to normal production (California State Water Pollution Control Board, 1964b). Other studies covered physical and chemical properties of the waters, phytoplankton, zooplankton, nekton, and the benthos (California State Water Pollution Control Board, 1964a, 1965a, 1965b; Hartman, 1955, 1960; and others). In some cases, ecological studies preceded installation of outfalls, one example being that of Turner *et al.* (1964). In one post-installation study it was claimed that the effects of pollution on fishes ranged from the development of a dull-colored, listless, flabby condition to lesions produced about the body and tumor-like sores about the mouth (Young, 1964).

There is one well-known study of the effect on the sea fisheries of the outflow of a polluted river. Carruthers summarized, in English, Kalle's interesting studies on the effect of the Thames River outflow on the commercial fisheries of the North Sea, specifically that part of the North Sea within which the river outflow was entrained. Quoting Carruthers: "Dr. Kalle states that the commercial catch is about double the corresponding catch made in the rest of the North Sea, in the English Channel, and in the Kattegat/Skagerrak region . . . and it is about 25 times the catch reckoned for the Baltic Sea as a whole." The increased catch was attributed to the fertilizing effect of polluted Thames River water (Carruthers, 1954). However, river discharges in general increase the ocean's productivity (Liebman, 1940; Riley, 1937).

The third major marine habitat is that represented by certain harbors and enclosed bays which have moderate to negligible changes in salinity and limited water circulation. These fall generally within Pritchard's definition of neutral or inverse estuaries (Pritchard, 1952). Included are most bays and harbors of southern California and certain lagoons of Texas and eastern Florida. In the southern California estuaries of this type, salinities remain quite uniformly high during most of the year; occasionally, however, heavy rains ensue, and floods descend upon the estuaries. Mass mortalities may result (MacGinitie, 1939). The papers that deal with the ecology of pollution in southern California cover benthic polychaetes, amphipods, other benthic invertebrates, and marine borers. They include papers by Reish (1955, 1956, 1957a, 1957b, 1959), Reish & Winter (1954), Barnard (1958), Barnard & Reish (1959), and Mohr (1953). On the coast of the Gulf of Mexico this environment is represented by certain Texas bays, for which Odum (1960) and Odum *et al.* (1963) reported on the relationship of primary productivity to pollution, a topic considered separately below. One other example of a study of this environment is that of Gilet (1960) in the harbors at Marseilles, France.

In general, this environment has the advantage of stable salinities, so that one important variable, salinity, can be largely ignored in studies of pollutional effects alone. On the other hand, in some areas there are sudden floods. If a flood occurs there is danger, of course, that the biotic indicator assemblage that may have adjusted to pollution for some time may be suddenly destroyed or severely modified.

The fourth major marine environment is represented by positive estuaries that may differ geomorphologically but which all share the common characteristic of net seaward flow of fresh water. The formerly polluted part of Biscayne Bay is in this category. Positive estuaries may be fjords, coastal plain estuaries, or bar-built estuaries such as northern Biscayne Bay. There have been bacteriological surveys in most coastal areas to determine safe areas for shellfishing (available from state health departments), and there have been many studies on the effects of oil and other pollutants on oysters (for references, see Baughman, 1948; Newell, 1959; Nelson, 1960; Ingram & Wastler, 1961; and the annual Proceedings of the National Shellfisheries Association), but there have been surprisingly few studies on the ecological effects of pollution in this environment. The most comprehensive, dealing with both the benthic and fouling organisms, are those of

Filice in San Francisco Bay (Filice 1954a, 1954b, 1958, 1959). Filice found three zones, which he called "barren," "marginal," and "clean," corresponding with three decreasing degrees of pollution. In the barren zone he found only two benthic species. Both were present in meager quantities. In the marginal zone, he found a few tolerant species thriving in large total quantities. In the clean zone, he found many species living in relatively small total quantities. The pollutant consisted of mixed domestic sewage and industrial wastes. He considered his three zones analogous with the freshwater zones designated "very polluted," "polluted," and "clean" by Patrick (1949). Filice reported that similar results were obtained by Volk (1907) and Schlienz (1923) in saline areas, and that Wilhelmi (1916) listed marine and estuarine indicator species.

There have been four reports on the fauna of intertidal muds of polluted British estuaries (Alexander *et al.*, 1936; Bassindale, 1938; Fraser, 1932; Stopford, 1951). Fraser's work is more pertinent to the Biscayne Bay studies than the work of the other authors. In a moderately polluted area he found what he considered to be a "*Macoma* community . . . but differing from the typical community, 'd', as described by Petersen (1918) in the excessive numbers of molluscs present, and in the absence of forms such as *Arenicola.*" He found also evidence of adverse pollutional effects in the dwarfing of *Cardium edule* and *Mytilus edulis.*

Dean & Haskin (1964) reported on the benthic repopulation of the river portion of the Raritan River estuary, New Jersey, following partial abatement of pollution. The pollution consisted of industrial wastes and domestic sewage in an area with maximum bottom salinities of 1 to 24.1 $^{0}/oo$. For three years following abatement, the number of fresh-water species increased steadily. The number of marine species increased for two years, then decreased slightly in the third year. Theirs is the first study of the effects of pollution abatement on the biota of an estuary, although there have been several studies of biotic responses to pollution abatement in fresh water (see Brinkhurst, 1965; Carpenter, 1924; and Laurie & Jones, 1938). Reish (1957a) reported on the repopulation of benthic organisms in a newly dredged portion of a polluted harbor; however, pollution remained unabated. In addition, Hawkes (1962) reviewed the British estuarine work cited above, and Behre (1963) discussed the suitability of many algal species as indicators of pollution in salinities of 2.2 to 18.5 $^{0}/oo$.

There have been many studies which treated special topics in pollution biology, topics which include indicator organisms, simplified biological indices of water quality, eutrophication, primary productivity, artificial fertilization of sea lochs, and the toxicity of wastes (including pesticides) to marine organisms. A representative sampling of the literature is presented here, with emphasis on marine work.

Due to wide differences in the adaptability of aquatic organisms to polluting substances, there is a universal tendency for a few tolerant species to survive and reach unusual abundance under conditions of moderate pollution. As heavy pollution is approached, the number of species and the number of individuals decline. Thus, both the association of species and the numerical abundance of their populations indicate pollution. In a practical program for monitoring the quality of water, the chief advantage of such species is that they indicate not only the conditions at the time of sampling but also the conditions that have prevailed for some time past. Their two chief disadvantages are the time required for sampling and analysis, particularly if taxonomic problems are involved, and the fact that the biota does not indicate the specific pollutants involved.

There have been many attempts to provide simplified biological indices of pollution that would substitute for full-scale biological surveys. One of the most successful is the "Catherwood diatometer" described by Patrick *et al.* (1954). It consists of an anchored, partially submerged microscope slide holder. From statistical analyses of the population densities and number of species that attach to the slides in the holder, inferences are drawn concerning water quality. The method has been used in two Texas bays (Hohn, 1959). Another method is that of King & Ball (1964), in which the ratio of the weights of aquatic insects (intolerant) to tubificid worms (tolerant) provides a useful index of pollution. The authors found that a ratio of less than 10 characterized polluted areas.

From extensive sampling in polluted and unpolluted harbors and bays in southern California, Reish (1957b and 1960) found that the benthic polychaete *Capitella capitata* is indicative of high organic wastes. He also used the organism in field toxicity tests. In one test, mature specimens from laboratory culture were held in small cages in the water. Survival over a period of time was checked. He found three levels of survival: one in which the animals lived several days but did not feed; another in which they fed but did not reproduce; and a third in which they both fed and reproduced. The three survival levels were correlated with concentrations of dissolved oxygen (Reish & Barnard, 1960). In another test, the fauna collected in suspended sediment-collecting bottles was correlated with the degree of pollution; the occurrence of *C. capitata* was particularly indicative (Reish, 1961).

Bellan (1964) confirmed Reish's conclusions on the indicator value of *C. capitata,* because the same species occurs in polluted muds in southern France. Bellan emphasized, however, that this polychaete is found strictly in very fine sediments, and that it does not tolerate any fraction of coarse material. He listed several species found in various degrees of pollution, together with their characteristic substrates. The only genus common to both his polluted habitats and Biscayne Bay is *Glycera.*

Just as artificial eutrophication of lakes is a frequent problem (Mackenthum

& Ingram, 1964; Mackenthum, 1965), so also certain estuaries suffer from the addition of allocthonous nutrients. In Moriches Bay and Great South Bay on Long Island, unusually dense phytoplankton blooms have received the attention of Ryther (1954) and Ryther *et al.* (1958). Ryther concluded that pollution from duck farms bordering Moriches Bay might have provided the organic nitrogen compounds necessary for the bloom. A cause of widespread destruction of oysters in these bays has been sought. Lackey (1960) cited the opinion of the late Thurlow C. Nelson that metabolites released by organisms such as *Eutreptia,* a green euglenid, may have an adverse effect on oysters and other marine organisms. Lackey found that recent organic fertilization (domestic sewage) often produces a bloom of *Eutreptia* in bays. In his experience, where there are repeated blooms of *Eutreptia,* oysters do not thrive. The most recent study from the Moriches Bay area was that of Barlow *et al.* (1963), who reported on the cycling of nutrients in a river that empties into the bay. In Raritan Bay as well, extremely dense populations of plankton are found. Jeffries (1962) found that these result from a combination of rich nutrient supplies, sluggish circulation, efficient regeneration of nutrients, and scarcity of macroscopic algae. A recent report summarized observations in Raritan, Barnegat (New Jersey), Moriches, and Great South bays (United States President's Science Advisory Committee, 1965).

Two reports from overseas, one from New Zealand and the other from Norway, are concerned with estuarine eutrophication. Wilkinson (1964) described the progressive increase in abundance of *Ulva* and *Enteromorpha* over a 30-year period in a New Zealand estuary. The periodic mortality and decomposition of the algae resulted in massive releases of hydrogen sulfide. Føyn (1960) reported that sludge from sewage effluents blankets the bottom in what were formerly rich shrimping grounds in the inner Oslofjord, Norway, and that only surface waters are now suitable for fish. Scientists at the Institute of Marine Biology, University of Oslo, are studying means to stop the eutrophication process.

H. T. Odum (1960) and H. T. Odum *et al.* (1963) reported that the metabolism of polluted bays is strikingly different from that of unpolluted ones, as would be expected, since organic pollution sets up an extremely heterotrophic situation in which production is much less than respiration, i.e., organic matter is used up much faster than it is produced (E. P. Odum, 1963). Another recent study of metabolism is that of Basu (1965) in India. Basu found that the productivity of a polluted estuary was negligible, while that of an unpolluted estuary ranged from 0.375 to 0.562 mg of carbon per m^2 per day, using the light and dark bottle sampling technique.

The sea loch fertilization experiments to enhance fisheries production during World War II and after are of special interest in pollution biology because they represent the only attempt so far to enrich a large marine area by design. The growth rate of flounders and plaice was accelerated through the increased production of phytoplankton, zooplankton, and bottom fauna where hydrographic conditions were favorable (Gross *et al.*, 1944; Gross, 1949a, 1949b).

Toxic effects of various substances on marine life have been reviewed by Butler & Springer (1963) and by Hood *et al.* (1960). Toxicity studies are essential in setting standards for water quality. Marine literature is as yet minute compared with the extensive freshwater literature (California State Water Quality Control Board, 1963).

Northern Biscayne Bay

Geology. Geomorphologically, northern Biscayne Bay is a bar-built estuary on a shoreline of low relief and shallow water (Hela *et al.*, 1957; Pritchard, 1952). In shallow undredged portions, the sediments consist of quartz sands, shell fragments and other invertebrate skeletal remains, small pieces of eroded limestone and coral, and some silt- and clay-sized particles. In deeper, dredged portions where there is little current, the sediment is a soft, gray to black ooze of predominantly fine fractions. The range of median grain diameter is 0.05 to 0.83 mm (McNulty, 1961). Some data for particle sizes appear also in Thorp (1935).

Various governmental agencies and private engineering firms have taken core borings to a depth of 30.5 m. Sandy, silty mud extends to a depth of 0.3 to 4.6 m; below this layer unconsolidated sand, muck, marl, and rotten shell are found in a stratum which is 3.1 to 6.1 m thick (Morrill & Olson, 1955). Quartz sands, called Pamlico Sands, originate in the Piedmont region of the Appalachians (Martens, 1935; Thorp, 1935). Miami Oolite, a fairly soft, porous, gray and white lime rock, lies below the unconsolidated material described above. It is usually not greater than 6.1 m in thickness, although a maximum thickness of 12.2 m has been found (Parker *et al.*, 1955). Below the Miami Oolite lies the Fort Thompson formation, a grayish-white to tan calcareous sandstone. It is about 24.4 m thick in the Miami area.

Miami Oolite has been dated recently at about 100,000 years (J. E. Hoffmeister, personal communication). This formation, intensively studied in recent years by Hoffmeister and his group, extends from Boca Raton to the lower Keys. In the Miami area it underlies and helps to form the Atlantic Coastal Ridge, which rises along the bay's western shore to an average height of eight feet (Davis, 1943). Its eroded bayward face forms a conspicuous feature along South Bayshore Drive, Coconut Grove. South of Miami, it floors Florida Bay and reappears above water level to form the lower Keys (Parker *et al.*, 1955).

During the Aftonian interglacial stage in the early Pleistocene, the ocean depth over south Florida exceeded 76 m. Since then the Floridian Plateau has emerged, and there have been successive sea-level changes in response to glaciations and deglaciations. In the Recent Epoch, the sea level has risen to near its present height with the melting back of the Wisconsin ice sheets. As recently as 5,000 B.C., relative sea level stood at five–perhaps eight–feet higher than at present, where it remained for 2,000 to 3,000 years. It was then that the Silver Bluff terrace of the Atlantic Coastal Ridge along South Bayshore Drive, Coconut Grove, was carved by wave erosion. Approximately 3,000 years ago, the relative sea level dropped to that of today. West of the Atlantic Coastal Ridge, fresh water gradually replaced salt and brackish water in what is now the Everglades. Some of the old tidal channels crossing the coastal ridge came into use as discharge channels for the accumulated water in the Everglades, bringing about the formation of short streams such as the Miami River (Parker *et al.*, 1955).

Hydrography. The hydrography of northern Biscayne Bay was described in detail by Hela *et al.* (1957). This part of the bay is positive, its main source of fresh water from land drainage being the Miami River. A man-made channel dredged to a depth of 9.2 m between the ocean and the mainland shore is the main connection with the sea. The original mangrove-swamp shores and nearby lowlands have been filled with material dredged from adjacent sandy bottoms. Several causeways and residential islands have been constructed in like manner. The result is a complex of channels, islands, and relatively deep borrow areas among which the bottom remains at its original depth. Water depths at mean low tide at all except four stations ranged from 0.9 to 2.7 m; at stations 7, 15, 17, and 18, depths were 5.1, 9.0, 4.5, and 3.9 m, respectively (Fig. 1). Semidiurnal tides of 0.76 m at the ocean and 0.61 m along the bay's western shore are typical. Tidal currents attain speeds of 1.6 m/sec at the ocean end of the ship channel and 0.62 m/sec at its landward end. Typical tidal currents paralleling the mainland and Miami Beach shores reach speeds of 0.4 to 0.62 m/sec. The annual range of water temperature is approximately 20° C to 30° C. Salinity averages 20 to 32.5 $^{o}/oo$ at all stations except three in the Miami River's mouth, where the average is 5 to 20 $^{o}/oo$.

Biology. The first ecological studies of northern Biscayne Bay were concerned with deterioration of submerged structures. Fouling and boring organisms and wood-inhabiting fungi were intensively studied in the decade following World War II. Joseph & Nichy (1955) reviewed this literature, from which several studies warrant special mention. Weiss (1948) listed all major species of fouling organisms and described their occurrence. More extensive studies, including work in the bay several miles south of Miami, were reported by Smith, Williams & Davis (1950). They described the seasonal qualitative and quantitative changes of fouling organisms and their local geographical distribution, plus pertinent conditions of the hydrography and plankton. These studies made possible the inclusion of Biscayne Bay in the digest on marine fouling assembled from worldwide sources by the Woods Hole staff (Woods Hole Oceanographic Institution, 1952). The

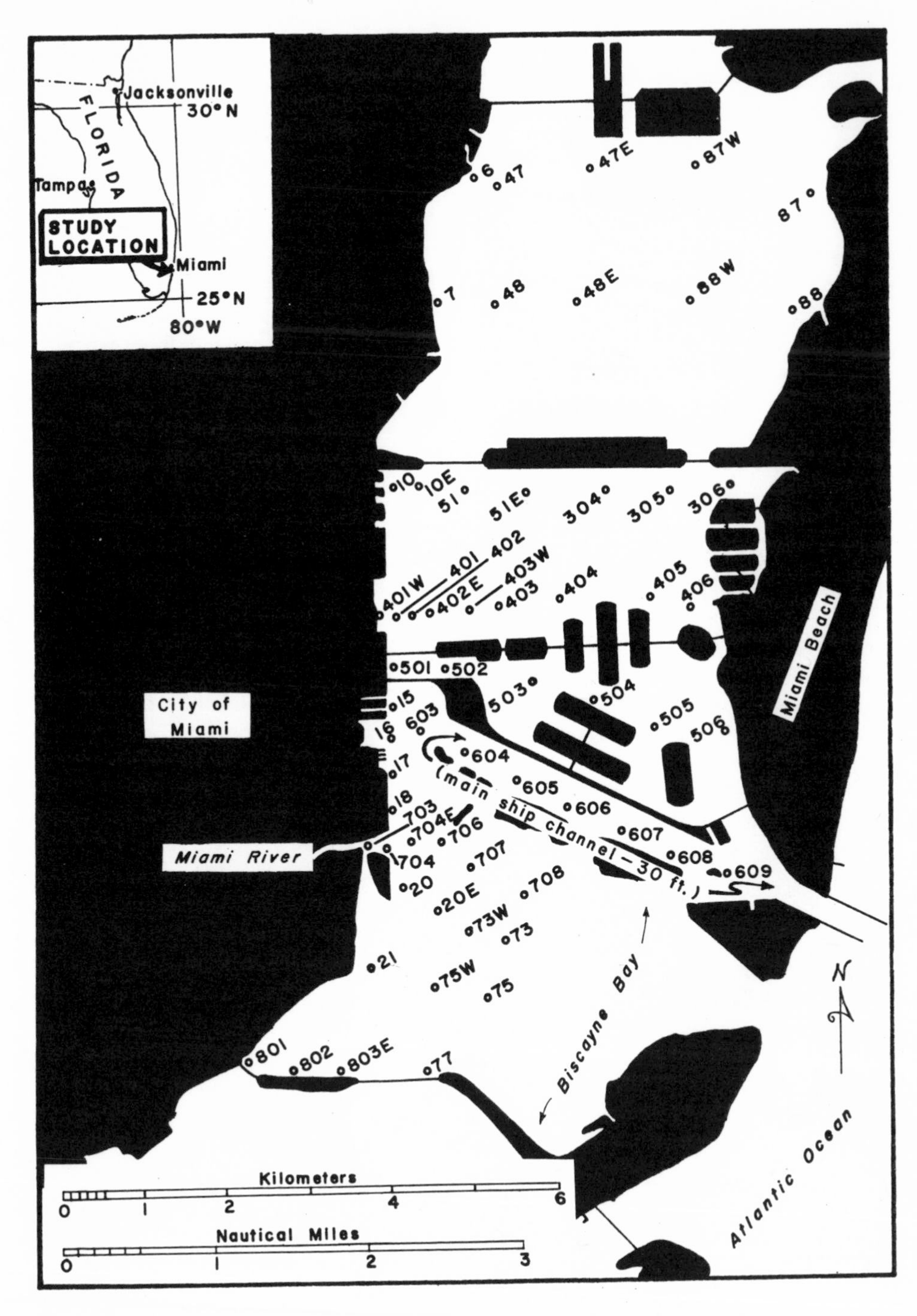

Fig. 1. Location of stations

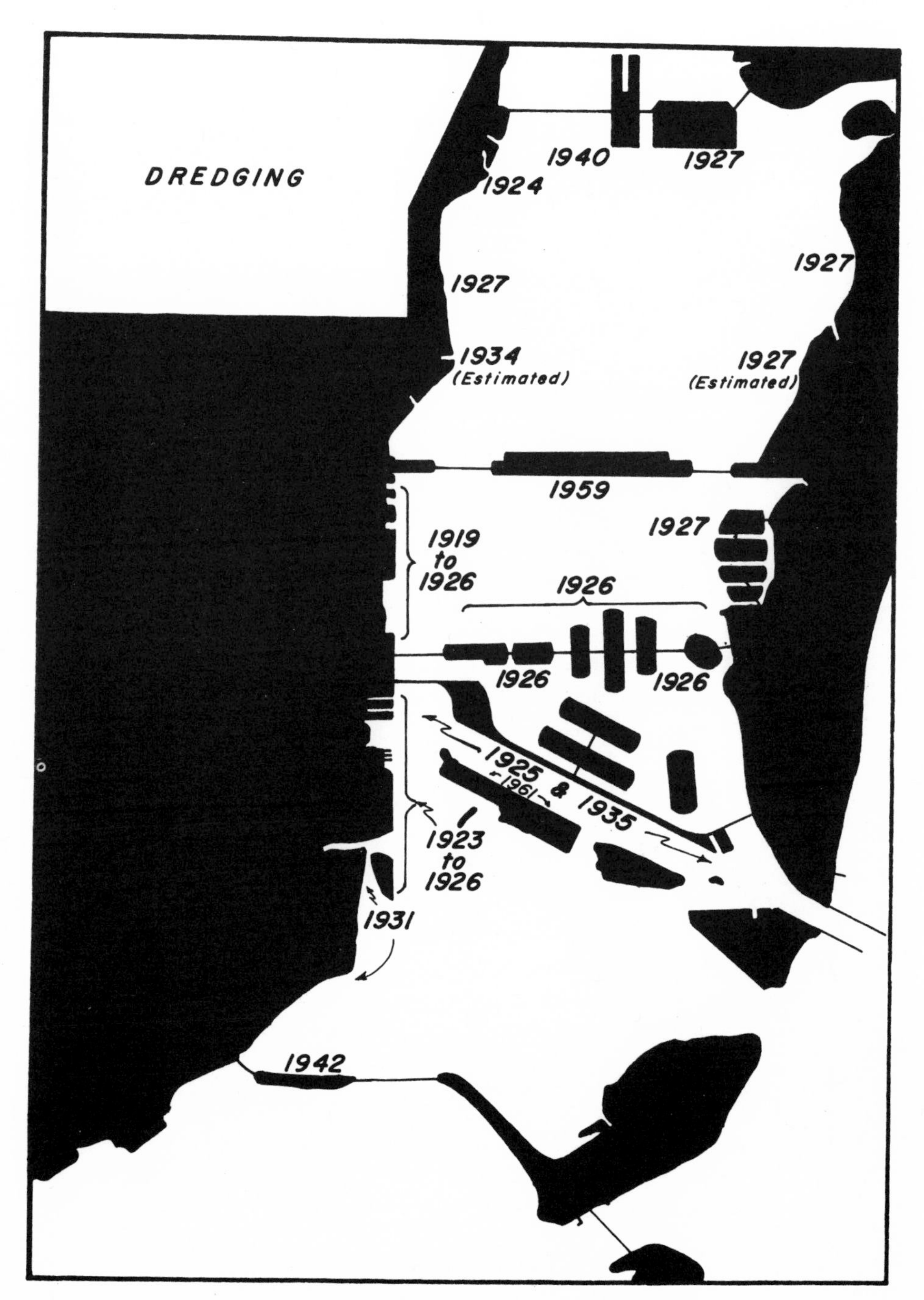

Fig. 2. Years of major dredge and fill projects
(*Source:* U.S. Army Corps of Engineers)

ecologies of marine borers and marine fungi were described by Greenfield (1952) and by Meyers (1953, 1954), respectively. Moore & Frue (1959) analyzed 12 years of data on the settlement, survival, and growth of three species of barnacles, correlating the data with temperature, lunar periodicity, river discharge, and the increase in human population.

Woodmansee (1949) described seasonal changes in the plankton some 17.7 km south of the Miami River, and Davis (1950) discussed plankton at two stations in the north bay area. Thomas *et al.* (1961) estimated the dry and wet weight of *Thalassia* washed ashore during Hurricane Donna. They concluded that despite tissue loss there was little permanent damage to the beds. D. R. Moore (1963) discussed the fauna associated with *Thalassia* in Biscayne Bay and the ecological factors that limit the distribution of *Thalassia.* Although concerned with an area well south of the polluted portion of Biscayne Bay, the ecological survey of Soldier Key by Voss & Voss (1955) is invaluable as a base study against which to gauge future environmental changes. The same can be said of the descriptions of bottom communities in areas south of formerly polluted parts of the bay (McNulty *et al.*, 1962a, 1962b), and the animal communities associated with *Thalassia* and *Diplanthera* (O'Gower & Wacasey, 1967).

Selected ecological effects of pollution before abatement were described by McNulty *et al.* (1960), and McNulty (1961), who treated the following subjects: distribution of coliform bacteria, chemical nutrients, zooplankton, sediments, and the distribution of benthic and fouling organisms. Preabatement studies of the benthos were conducted at 78 stations. Sixty of the original 78 stations were selected for postabatement studies of the benthos; their locations are shown in Figure 1. The postabatement studies have included dissolved phosphate-phosphorus, volumes of plankton, and fouling organisms at 13, 11, and 10 stations, respectively. An additional paper on pollution is that of Lynn & Yang (1960) in which a wet-oxidation method for the determination of the organic carbon of sediments was proposed.

Dredging and Filling. With the completion of Henry M. Flagler's railroad to Miami and the arrival of the first passenger train in 1896, a demand was created for filled land and a deepwater seaport. The first channel to the sea was dredged across the shallows between the mouth of the Miami River and Cape Florida in the late 1890's. By 1915, there was a channel 5.5 m deep and 46 m wide from Government Cut to the west shore of the bay (Hollingsworth, 1936). In 1925, the channel was deepened to 7.6 m and widened to 92 and 152 m (United States Board of Engineers for Rivers and Harbors, 1922). In the boom years following World War I, Miami and Miami Beach "experienced a burst of growth and progress well-nigh incredible, even to those who knew it best and were most enthusiastic as to its future" (Munroe & Gilpin, 1930). As the dates in Figure 2 show, most of the dredging and filling took place between the wars, so that northern Biscayne Bay at midcentury bore little resemblance to its original aspect of the early 1900's. Construction has continued in recent years with the Julia Tuttle Causeway (1959) and the county port facilities (1961).

Pollution. As Miami's population increased from 1,681 to 249,276 persons between 1900 and 1950, much of its domestic sewage was discharged to the Miami River or directly into Biscayne Bay (see Dade County, 1960; Milliken, 1949; Minkin, 1949). A point was reached finally at which pollution became an anathema to local residents as it progressively threatened tourism, a mainstay of the local economy. In response to public demand for abatement, a new disposal plant was constructed by the City of Miami on Virginia Key. Public outfalls to the river and bay were sealed, and the plant began treatment of some 136 to 227 million liters per day of domestic sewage in late 1955. Thus, quite suddenly, an important ecological factor in the north bay environment–heavy pollution from domestic sewage–was virtually eliminated.

By 1959, median coliform counts had dropped to below 1,000 MPN throughout the bay except in the Miami River and for 1.5 km due north of its mouth, and–aberrantly–in the main ship channel 4 km due east of the river mouth. A comparison of the data for 1949 with those for 1959 shows that median coliform counts in the river near its mouth had dropped from 110,000 to 33,000 MPN; off Bayfront Park (0.5 km due north of the river's mouth) from 24,000 to 7,000 MPN; and off Pier 2 (1.5 km due north of the river's mouth) from 11 million to 1,300 MPN (Dade County Health Department, personal communication). Major sources of continuing pollution were several private domestic outfalls in the river, yachts in the river and in a marina 1 km due north of the river's mouth, and ships in the Port of Miami (1.0 to 1.8 km due north of the river's mouth). Industrial wastes, consisting mainly of effluents discharged into the Miami River by laundries and light industries, have not been a serious problem in the bay. Figure 3 shows the distribution of pollution in 1949.

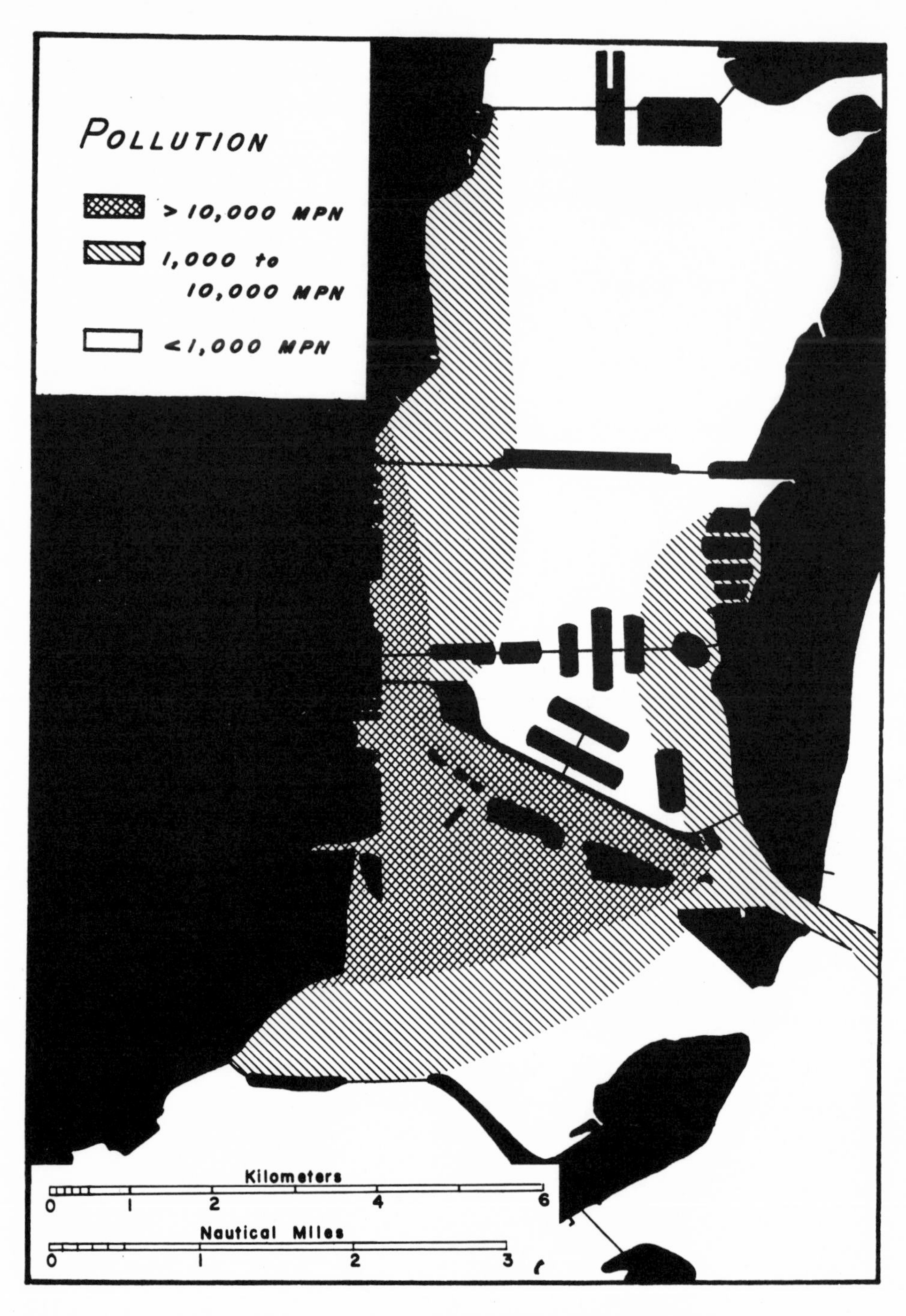

Fig. 3. Distribution of pollution (mean coliform bacteria per 100 ml–after Minkin, 1949)

Methods

Benthos

Since the purpose of this study was to compare the benthos before and after pollution abatement, it was particularly important to use the same sampling techniques in 1960 that were used in the original study. The same Petersen grab was used, mounted from a small boat in exactly the same way as before. Sixty of the original 76 stations were resampled. Sixteen of the original stations were omitted because they were located outside the area of principal concern. Three grab samples were taken at each station, each sample representing 0.074 m^2 of bottom, the sampling area of the grab. Sieving was done in the field with a box fitted with a 1-mm-mesh-opening screen. The sieve box with its contained sample was lifted over the boat's side, immersed, and agitated until all materials small enough to wash through had disappeared. The unsorted samples were then bottled and preserved in 10 percent formalin and returned to the laboratory for sorting, identification, and enumeration of fauna.

Inorganic Phosphate-Phosphorus

For the sake of greater efficiency and accuracy, a superior method of inorganic phosphate-phosphorus determination was used in 1960-61, as compared with that used in 1956. In 1960-61, the method used was that of Murphy & Riley (1958), a modification of the method of Greenfield & Kalber (1954). In the earlier work, the method was that of Robinson & Thompson (1948). Sampling was monthly from January to July, 1961, the same part of the year as in the original work.

Plankton

Monthly plankton samples were obtained between January and July, 1961, with a Clarke-Bumpus sampler and No. 2 net. The seasons of sampling were the same as in the original work, and the sampler was the same. Plankton volumes were determined by draining a sample on a piece of No. 2 net and measuring the displacement volume in a small graduated cylinder. Calculations of plankton volume per m^3 were made, using the average calibration factor of 4.1 liters per revolution from the data in Clarke & Bumpus (1940).

Fouling Organisms

Glass panels, measuring 10.2 x 12.7 cm, in anchored wooden frames exposed at depths of 1.5 to 2.4 m below MLW, were examined monthly between January and June, 1961, to compare the numerical abundance of barnacles and amphipod tubes and the total displacement volume of all fouling organisms with like data obtained in the original study. Station locations and seasons of exposure were the same in both studies. Enumeration of barnacles and amphipod tubes was done by counting individuals within twenty 4-cm^2 frames located at random over the glass panels. Displacement volumes were determined by scraping the panels clean and measuring the volume of all fouling organisms in a graduated cylinder.

Results

Benthos

Table 1 compares the species taken with the Petersen grab in both 1956 and 1960. Tables 2-A through 2-F consist of an array of the 1960 data, in exactly the same format as that for the 1956 data (McNulty, 1961), and are the basis for the analyses that follow.

The author described two bottom communities of northern Biscayne Bay before abatement of pollution. One community was found mainly in polluted areas (greater than 10,000 MPN), while the other characterized less polluted areas (less than 10,000 MPN) (McNulty, 1961).

The pollution-tolerant community of 1956 was characterized mainly by the red algae *Gracilaria blodgettii* and *Agardhiella tenera,* plus the tubicolous errant polychaete, *Diopatra cuprea.* In 1960, absolutely no red algae, but some *Diopatra cuprea,* were taken within the boundaries set for this community in earlier work.

TABLE 1

Major Species of Benthic Macroorganisms Taken with the Petersen Grab in Biscayne Bay, North of Rickenbacker Causeway, Miami, Florida, July-August, 1956 (polluted) and November-December, 1960 (unpolluted)

Species	1956	1960
SPERMATOPHYTA		
Diplanthera wrightii (Ascherson)	x[1]	x
Halophila baillonis Ascherson	x	x
Syringodium filiforme Kützing	x	x
Thalassia testudinum König	x	x

Table 1, continued

Species	1956	1960
CHLOROPHYCEAE		
Bryopsis hypnoides Lamouroux	x	–
Caulerpa mexicana (Sonder) J. Agardh	x	–
Caulerpa prolifera (Foersskal) Lamouroux	x	–
Chaetomorpha brachygona Harvey	x	–
Cladophora gracilis (Griffiths *ex* Harvey) Kützing	x	–
Enteromorpha prolifera (Müller) J. Agardh	x	–
Ulva lactuca Linnaeus	x	–
RHODOPHYCEAE		
Acanthophora spicifera (Vahl) Bǿrgesen	x	x
Agardhiella tenera (J. Agardh) Schmitz	x	x
Ceramium tenuissimum (Lyngbye) J. Agardh	x	–
Cryptonemia luxurians (Mertens) J. Agardh	x	–
Gracilaria blodgettii Harvey	x	x
Hypnea cervicornis J. Agardh	x	x
H. cornuta (Lamouroux) J. Agardh	x	–
Laurencia obtusa (Hudson) Lamouroux	x	–
Polysiphonia subtilissima Montagne	x	–
Wrangelia bicuspidata Bǿrgesen	x	–
PHAEOPHYCEAE		
Dictyota dichotoma (Hudson) Lamouroux	x	–
XANTHOPHYCEAE		
Vaucheria sp.	x	–
OPHIUROIDEA		
Amphiodia pulchella (Lyman)	x	–
Amphioplus abditus (Verrill)	x	x
A. coniortodes H. L. Clark	x	x
Amphipholis gracillima (Stimpson)	x	x
Ophionephthys limicola Lütken	x	x
POLYCHAETA		
Arabella iricolor (Montagu)	–	x
Armandia agilis (Andrews)	–	x
Branchiomma nigromaculata (Baird)	x	–
Capitomastus aciculatus Hartman	–	x

Table 1, continued

Species	1956	1960
Chaetopterus variopedatus (Renier)	–	x
Cistenides gouldii Verrill	x	x
Diopatra cuprea (Bosc)	x	x
Glycera americana Leidy	–	x
G. dibranchiata Ehlers	–	x
G. tesselata Grube	–	x
Glycera sp.[2]	x	x
Lumbrineris maculata (Treadwell)	–	x
Lumbrineris sp.	x	x
Lysidice ninetta Audouin & Milne Edwards	–	x
Maldane sarsi Malmgren	x	–
Marphysa sanguinea (Montagu)	–	x
Naineris setosa (Verrill)	x	–
Onuphis magna (Andrews)	x	–
Onuphis sp.	–	x
Owenia fusiformis delle Chiaje	x	x
Pista cristata (Müller)	x	x
Pista sp.	x	–
Polydora sp.	x	–
Scoloplos (Leodamas) rubra (Webster)	–	x
Piromis roberti (Hartman)	x	x
Spiochaetopterus oculatus Webster	x	x
Sthenelais sp.	x	x
Terebellides stroemi Sars	x	x
Sabellaria sp.	–	x
SIPUNCULIDA[3]		
Golfingia sp.	–	x
Phascolion sp.	–	x
CRUSTACEA[4]		
Alpheus floridanus Kingsley	–	x
Corophium acherusicum Costa	x	–
Eucratopsis crassimanus (Dana)	x	x
Grubia filosa (Savigny)	x	–
Grubia sp.	x	–
Pagurus annulipes (Stimpson)	–	x
Penaeus duorarum Burkenroad	–	x
Processa sp.	–	x
Trachypenaeus constrictus (Stimpson)	–	x

Table 1, continued

Species	1956	1960
GASTROPODA		
Anachis avara (Say)	x	–
Bulla occidentalis A. Adams	x	–
Bulla striata Bruguière	–	x
Conus floridanus Gabb	x	–
Conus stearnsii Conrad	x	–
Epitonium rupicolum Kurtz	–	x
Haminoea antillarum guadalupensis Sowerby	x	–
Mitrella lunata Say	x	–
Nassarius vibex (Say)	x	x
Neritina virginea (Linnaeus)	x	–
Olivella (Minioliva) perplexa Olsson		x
Olivella sp.	–	x
Turbo castaneus Gmelin	x	–
LAMELLIBRANCHIATA		
Atrina rigida (Solander)	x	–
Botula castanea (Say)	x	–
Cardita floridana (Conrad)	x	x
Chione cancellata (Linnaeus)	x	x
Codakia costata d'Orbigny	–	x
C. orbicularis (Linnaeus)	x	–
Corbula caribaea d'Orbigny	–	x
C. swiftiana C. B. Adams	–	x
Cumingia tellinoides (Conrad)	x	x
Cyclinella tenuis (Récluz)	x	x
Dosinia elegans (Conrad)	x	x
Laevicardium mortoni (Conrad)	x	x
Lima pellucida C. B. Adams	x	–
Lioberus castaneus Say	–	x
Lucina multilineata (Tuomey & Holmes)	x	–
Macoma tenta Say	–	x
Macoma sp.	x	–
Mactra fragilis Gmelin	x	–
Modiolus papyria (Conrad)	x	–
Tagelus divisus (Spengler)	x	x
Tellina alternata Say	x	–
T. lineata Turton	–	x
T. martinicensis d'Orbigny	–	x

Table 1, continued

Species	1956	1960
T. versicolor Cozzens	x	x
Tellina sp.	x	—
Nucula proxima Say	—	x
Pitar fulminata (Menke)	—	x
Trachycardium egmontianum (Shuttleworth)	x	—
Trachycardium muricatum (Linnaeus)	x	—

1 "x" means species was taken.
2 Glycerids not identified to species in 1956.
3 Unidentified sipunculids were present in 1956.
4 Other crustaceans, unidentified, were present in 1956.

Tables 2-A – F

Species and Quantities of Plants and Animals Taken in Three Petersen Grab Samples per Station (total area 0.222 m^2 per station), from 17 November through 8 December, 1960, in Biscayne Bay, Florida (Quantities of Plants are in Milliliters Displacement Volume; Quantities of Animals are in Numbers of Individuals)

Table 2-A

	Station number									
Species	6	47	47E	87W	87	7	48	48E	88W	88
SPERMATOPHYTA										
Diplanthera wrightii	–	–	–	–	–	–	–	–	10	100
Halophila baillonis	–	–	–	–	–	–	–	–	–	5
Syringodium filiforme	–	–	–	–	–	–	tr[1]	175	30	–
Thalassia testudinum	–	–	–	–	–	–	5	–	55	–
RHODOPHYCEAE										
Acanthophora spicifera	–	–	–	–	–	–	tr	tr	10	–
Agardhiella tenera	–	–	–	–	–	–	–	3	tr	–
Gracilaria blodgettii	–	–	–	–	–	–	–	tr	tr	–
Hypnea cervicornis	–	–	–	–	–	–	–	–	3	–
POLYCHAETA										
Capitomastus aciculatus	–	–	3	–	1	–	–	–	–	–
Cistenides gouldii	–	1	–	1	1	2	3	1	–	–
Diopatra cuprea	1	2	–	–	–	–	2	10	–	–
Glycera americana	–	–	–	–	–	2	–	–	–	1
G. dibranchiata	1	1	1	12	1	1	–	–	1	–
Marphysa sanguinea	–	–	–	–	–	–	–	–	1	–

Table 2-A, continued

Species	6	47	47E	87W	87	7	48	48E	88W	88
	Station number									
Owenia fusiformis	–	–	–	–	–	–	1	–	2	2
Pista cristata	–	–	–	–	–	–	3	–	–	–
Sabellaria sp.	–	–	–	–	–	–	2	–	–	–
Scoloplos (Leodamas) rubra	–	–	–	–	–	1	–	–	–	–
Semiodera roberti	–	–	–	–	–	–	2	–	–	–
Spiochaetopterus oculatus	–	–	–	–	–	21	–	–	–	–
Unidentified polychaetes	–	–	–	–	–	–	–	2	1	–
CRUSTACEA										
Unidentified caridean	–	–	–	–	–	–	1	1	–	–
Unidentified hermit crab	–	–	–	–	–	–	3	–	–	–
LAMELLIBRANCHIATA										
Cardita floridana	–	–	–	–	–	–	3	–	–	–
Chione cancellata	–	–	–	–	2	–	3	6	2	1
Corbula caribaea	–	–	–	2	–	–	–	–	–	–
C. swiftiana	–	–	–	7	35	–	1	–	–	–
Cumingia tellinoides	–	–	–	–	–	–	1	–	–	–
Cyclinella tenuis	–	–	–	1	–	4	–	–	–	–
Diplodonta nucleiformis	–	–	–	–	–	–	1	–	–	–
Laevicardium mortoni	–	–	–	–	1	–	1	1	–	1
Macoma tenta	–	–	–	–	–	–	1	–	–	–
Mactra fragilis	–	–	–	–	–	1	–	–	–	–
Tagelus divisus	–	3	2	3	20	8	1	–	–	–
Tellina martinicensis	–	–	–	–	–	–	–	1	–	–
Tellina versicolor	–	–	–	–	1	–	2	–	–	–
OPHIUROIDEA										
Amphioplus abditus	–	–	–	–	–	–	2	2	4	1
Amphioplus coniortodes	–	–	–	–	–	1	–	–	–	–
Amphipholis gracillima	–	–	–	–	–	1	–	–	–	–

1 "tr" means trace

TABLE 2-B

Species	10	10E	51	51E	304	305	306	401W	401	402	402E
	Station number										
SPERMATOPHYTA											
Halophila baillonis	–	31	–	–	–	–	–	–	–	–	–
POLYCHAETA											
Cistenides gouldii	–	–	–	2	1	1	–	5	–	1	–
Diopatra cuprea	–	1	1	–	–	–	–	–	–	–	2
Glycera americana	–	–	–	1	1	–	–	–	2	–	2
G. dibranchiata	–	2	1	–	–	–	–	–	–	2	2
G. tesselata (?)	–	–	–	–	–	–	–	–	–	1	–

Table 2-B, continued

Species	Station number										
	10	10E	51	51E	304	305	306	401W	401	402	402E
Glycera sp.	–	–	–	–	–	–	–	–	–	2	1
Lumbrineris sp.	–	1	–	–	–	–	–	–	–	2	–
Onuphis sp.	–	1	–	–	–	–	–	–	–	–	–
Owenia fusiformis	–	–	–	–	–	–	–	2	3	6	–
Scoloplos (Leodamas) rubra	–	–	–	–	–	1	–	–	1	–	1
Sthenelais sp.	–	–	–	3	–	–	–	–	–	–	–
Unidentified polychaetes	–	4	1	3	–	–	–	1	–	–	–
SIPUNCULIDA											
Golfingia sp.	–	–	–	–	–	–	–	12	–	–	1
Phascolion sp.	–	–	–	–	–	–	–	1	–	–	–
Unidentified sipunculids	4	4	33	–	–	–	–	–	–	2	1
CRUSTACEA											
Alpheus floridanus	–	–	–	–	–	–	–	–	1	–	–
Eucratopsis crassimanus	–	1	–	–	1	–	–	–	–	–	–
Penaeus duorarum	–	–	–	1	–	–	–	1	–	–	–
Trachypenaeus constrictus	–	–	–	–	–	–	–	–	1	–	–
GASTROPODA											
Bulla striata	–	–	3	–	–	–	–	–	–	–	–
Epitonium rupicolum	–	–	–	–	–	–	–	1	–	–	–
Nassarius vibex	1	–	–	–	–	–	–	–	–	–	–
Olivella (Minioliva) perplexa	–	–	–	–	–	–	–	1	–	–	–
LAMELLIBRANCHIATA											
Cardita floridana	–	–	–	–	–	–	–	1	–	–	–
Chione cancellata	–	6	5	7	2	5	–	–	–	2	8
Codakia costata	–	6	13	2	–	–	3	2	–	25	9
Corbula swiftiana	–	–	–	–	1	–	–	–	–	–	–
Cyclinella tenuis	1	–	1	8	1	2	2	–	–	–	1
Dosinia elegans	–	–	–	–	–	–	–	–	1	–	–
Laevicardium mortoni	–	2	–	–	–	1	–	–	–	2	3
Lioberus castaneus	–	–	–	–	–	–	–	–	–	–	2
Macoma tenta	–	–	5	9	2	4	1	–	–	–	–
Nucula proxima	–	–	1	2	–	1	2	–	–	14	13
Tagelus divisus	48	10	–	9	50	17	7	9	–	3	7
Tellina martinicensis	–	–	–	–	–	1	–	–	–	–	–
T. versicolor	–	5	6	2	2	5	1	4	–	1	1
OPHIUROIDEA											
Amphioplus abditus	–	–	–	–	–	–	–	–	–	–	1
Ophionephthys limicola	–	–	–	–	–	–	–	–	–	–	1
ASCIDIACEA											
Unidentified ascidian	–	2	–	–	–	–	–	–	–	–	–

TABLE 2-C

Species	403W	403	404	405	406	501	502	503	504	505	506
					Station number						
SPERMATOPHYTA											
Syringodium filiforme	–	–	–	–	tr[1]	–	–	–	–	–	–
Diplanthera wrightii	–	–	–	–	57	–	–	–	–	–	–
NEMERTEA											
Unidentified nemertean	3	–	–	–	–	–	–	–	–	–	–
POLYCHAETA											
Cistenides gouldii	–	–	–	–	–	–	–	–	–	1	–
Diopatra cuprea	–	–	–	–	3	–	–	–	–	–	–
Glycera americana	–	3	–	1	1	2	–	–	–	–	–
G. dibranchiata	–	–	–	–	–	–	–	–	1	–	1
Lumbrineris sp.	–	–	–	–	–	1	–	–	1	6	1
Scoloplos (Leodamas) rubra	1	2	–	–	3	–	–	–	–	–	3
Sthenelais sp.	1	–	–	–	–	–	–	–	–	–	2
Unidentified polychaetes	–	–	–	–	1	1	2	–	5	–	12
SIPUNCULIDA											
Unidentified sipunculid	–	–	–	–	1	–	–	–	–	–	–
CRUSTACEA											
Pagurus annulipes	–	–	–	–	1	–	–	–	–	–	–
Processa sp.	–	–	–	–	2	–	–	–	–	–	–
GASTROPODA											
Olivella sp.	1	–	–	–	–	–	–	–	–	–	–
LAMELLIBRANCHIATA											
Chione cancellata	1	–	–	–	–	–	1	–	–	–	–
Codakia costata	–	–	–	–	4	7	–	–	–	–	–
Cyclinella tenuis	2	1	1	1	2	1	3	–	2	–	–
Laevicardium mortoni	–	–	–	–	1	–	1	–	–	–	–
Macoma tenta	–	3	1	2	–	–	–	–	1	1	–
Tagelus divisus	5	6	17	15	9	–	2	9	6	5	4
Tellina martinicensis	2	1	–	–	–	–	–	–	1	1	–
T. versicolor	3	4	1	6	3	–	3	–	1	1	1
Nucula proxima	–	–	–	–	–	–	8	–	–	–	–
OPHIUROIDEA											
Amphioplus abditus	1	1	2	–	2	–	3	–	1	–	–
A. coniortodes	–	–	–	–	–	–	–	–	–	–	1
Amphipholis gracillima	–	–	1	–	1	–	1	1	–	–	–
Ophionephthys limicola	3	1	1	–	–	6	1	–	1	3	1

1. "tr" means trace

TABLE 2-D

Species	15	16	17	18	603	604	605	606	607	608	609
					Station number						
SPERMATOPHYTA											
Diplanthera wrightii	–	–	–	–	–	–	–	–	–	15	120
Thalassia testudinum	–	–	–	–	–	–	–	–	–	–	30
POLYCHAETA											
Cistenides gouldii	–	2	–	–	–	–	–	1	–	–	2
Diopatra cuprea	–	–	–	–	–	–	–	–	–	–	2
Glycera americana	–	–	–	1	–	1	–	–	–	1	1
G. dibranchiata	–	1	–	–	–	–	–	–	–	–	–
Lumbrineris sp.	1	–	–	–	–	–	1	1	–	–	–
Lysidice ninetta	–	–	1	–	–	–	–	–	–	–	–
Owenia fusiformis	–	4	–	–	–	4	–	–	3	1	3
Scoloplos (Leodamas) rubra	–	1	–	–	–	–	1	–	–	–	–
Semiodera roberti	–	–	–	–	–	–	–	–	–	–	1
Unidentified polychaetes	1	2	–	–	–	1	1	2	–	–	1
SIPUNCULIDA											
Unidentified sipunculids	–	–	–	–	–	6	–	–	–	–	–
CRUSTACEA											
Eucratopsis crassimanus	–	–	–	–	–	1	–	–	–	–	–
GASTROPODA											
Bulla striata	–	2	–	–	–	–	–	–	–	–	2
LAMELLIBRANCHIATA											
Chione cancellata	–	1	–	–	–	3	2	–	1	2	–
Codakia costata	–	8	–	–	–	3	–	–	–	–	–
Cyclinella tenuis	–	–	–	–	–	–	1	–	–	–	–
Laevicardium mortoni	–	1	–	–	–	–	1	–	–	–	–
Macoma tenta	–	–	–	–	–	–	–	–	–	1	–
Tagelus divisus	–	3	3	–	–	1	1	9	–	1	1
Tellina versicolor	–	1	–	–	–	1	–	–	–	–	2
OPHIUROIDEA											
Amphioplus abditus	–	–	–	–	–	1	–	–	–	–	1
Amphipholis gracillima	–	–	–	–	–	–	1	1	–	–	–
Ophionephthys limicola	–	–	–	–	4	–	–	1	–	–	–

TABLE 2-E

Species	Station number 703	704	704E	706	707	708	20	20E	73W	73
SPERMATOPHYTA										
Halophila baillonis	–	–	tr[1]	–	–	–	–	–	tr	–
RHODOPHYCEAE										
Acanthophora spicifera	–	–	–	–	–	–	–	5	tr	–
POLYCHAETA										
Chaetopterus variopedatus	–	–	–	–	1	–	–	–	–	–
Cistenides gouldii	–	–	–	–	–	–	–	1	1	–
Diopatra cuprea	2	–	–	–	–	–	–	–	–	–
Glycera americana	–	–	3	1	–	–	–	1	–	1
G. dibranchiata	–	–	–	1	–	–	2	3	–	–
Lumbrineris maculata	–	–	1	–	4	1	–	1	–	1
Owenia fusiformis	–	1	3	3	–	1	2	3	2	5
Scoloplos (Leodamas) rubra	–	–	1	–	–	–	–	–	–	–
Terebellides stroemi	–	–	–	–	–	–	–	–	–	2
Unidentified polychaetes	1	1	2	1	–	–	–	–	–	–
CRUSTACEA										
Alpheus floridanus	–	–	–	–	4	1	–	–	–	–
Unidentified carideans	–	–	–	–	–	–	–	–	2	–
LAMELLIBRANCHIATA										
Chione cancellata	–	1	7	2	–	4	3	4	2	4
Codakia costata	–	–	35	27	–	4	41	–	11	22
Cyclinella tenuis	–	–	1	1	1	4	1	–	–	3
Dosinia elegans	–	–	–	–	1	–	–	–	–	–
Laevicardium mortoni	–	–	–	1	–	–	1	1	1	1
Lucina sp.	–	–	–	–	–	1	–	–	–	–
Macoma tenta	–	1	–	–	–	–	–	–	–	1
Tagelus divisus	1	–	1	3	2	6	1	–	2	–
Tellina lineata	–	–	–	–	–	–	–	–	–	1
T. martinicensis	–	–	–	–	–	–	–	–	–	1
T. versicolor	–	–	1	3	1	–	2	–	–	1
Nucula proxima	–	–	–	1	–	–	7	–	–	–
OPHIUROIDEA										
Amphioplus abditus	–	–	–	–	–	1	1	–	1	2
A. coniortodes	–	–	–	–	–	–	–	–	–	2
Ophionephthys limicola	–	–	1	1	3	3	–	–	–	1

[1]"tr" means trace.

TABLE 2-F

Species	Station number						
	21	75W	75	801	802	803E	77
NEMERTEA							
Unidentified nemertean	–	1	–	–	–	–	–
POLYCHAETA							
Arabella iricolor	–	–	–	1	–	–	–
Armandia agilis	–	–	1	–	–	–	–
Chaetopterus variopedatus	1	–	–	–	–	–	–
Diopatra cuprea	1	–	–	8	–	4	1
Glycera dibranchiata	3	–	–	5	1	–	1
Lumbrineris sp.	–	1	–	5	–	–	–
Lumbrineris sp.	1	–	2	1	1	–	–
Owenia fusiformis	5	–	–	17	–	1	–
Sabellaria sp.	–	–	–	5	–	–	–
Unidentified polychaetes	–	–	–	1	1	1	1
LAMELLIBRANCHIATA							
Chione cancellata	–	2	–	51	3	1	1
Codakia costata	3	–	–	31	1	–	4
Cyclinella tenuis	–	–	–	1	–	–	–
Dosinia elegans	–	–	–	–	–	–	1
Laevicardium mortoni	–	–	–	–	–	1	–
Lioberus castaneus	2	–	–	5	–	–	–
Nucula proxima	1	–	–	1	–	–	–
Pitar fulminata	–	–	–	–	–	–	1
Tagelus divisus	–	1	–	3	–	–	–
Tellina martinicensis	–	–	–	–	1	–	–
T. versicolor	–	–	–	1	–	–	–
OPHIUROIDEA							
Amphioplus abditus	3	–	1	1	–	1	–
A. coniortodes	–	–	–	–	7	–	–
Ophionephthys limicola	–	–	–	1	4	–	–
ASCIDIACEA							
Unidentified ascidian	–	–	–	1	–	–	–

Also, in 1960 a group of five species originally assigned to this community was not taken: *Branchiomma nigromaculata, Dictyota dichotoma, Haminoea antillarum, Laurencia obtusa,* and *Ulva lactuca.* In all, 8 of 19 species (42 percent) selected as especially characteristic of polluted areas in 1956 were absent from samples in 1960. The conclusion is that the pollution-tolerant community of 1956 had simply ceased to exist as a describable entity in 1960, even though 58 percent of its individual species were still taken in formerly polluted areas.

However, diversion of sewage was not the only important ecological change during the years between sampling. Construction of the new Julia Tuttle Causeway just north of stations 10 through 306, and the start of construction of new port facilities on spoil islands next to the ship channel—specifically, hydraulic dredging required for their construction—were major disruptive influences. Also, Hurricane Donna, on October 10, 1960, brought sustained easterly winds of 55-63 mph with gusts of 80-82 mph.

The influence of these and possibly other disruptive factors can be gauged by what happened to the bottom community formerly found in weakly polluted areas, because if pollution abatement had been the only important ecological change between samplings, then the biota of such areas should have changed little or not at all. (Actually this is an oversimplification because of possible cyclic changes in population and purely random fluctuations resulting from spawning irregularities, predator-prey imbalances, and other unknowns.) The bottom community of weakly polluted areas in 1956 (less than 10,000 MPN) was characterized primarily by the spermatophytes *Halophila baillonis* and *Diplanthera wrightii,* plus the ophiuran *Amphioplus abditus.* In 1960, this plant-animal association was found at only 5 of 23 stations where formerly found. It persisted at stations approximately one mile or more distant from hydraulic dredging sites, in areas which hydrographically would receive a minimum of turbidity and sedimentation effects from dredging, and in areas well protected from effects of strong easterly winds. As for the remaining 18 stations that formerly supported this pollution-intolerant community, only one continued to support spermatophytes in 1960. This was not the case with the ophiuran, however, which continued to flourish at scattered stations throughout the area. In general, the distribution of animal species of this community changed very little, and the level of abundance of individuals per m^2 remained about the same. These facts are interpreted to mean that dredging, and possibly Hurricane Donna, caused widespread damage to attached vegetation, but had little effect on benthic infauna.

In the analysis that follows in the next section, both of the communities described above are combined, and redescribed as "the hard-bottom community." The community is clearly identifiable with a recently described clean-water hard-bottom community to the south in Biscayne Bay. The affinities are discussed and the effects of pollution upon the community are described in detail.

Two other ecological zones were found in 1956: a zone of essentially no life, and a zone characterized by little or no attached vegetation and presence of the ophiuran *Ophionephthys limicola.* Both zones occupy former dredge areas that are deeper than surrounding undredged areas, and both have soft sediments. Both zones were clearly identifiable in 1960. They are discussed below under the heading, "The Soft-Bottom Community."

Analysis of Benthos

The purpose of the analysis is to describe changes in normal bottom communities attributable to pollution alone. Description of normal communities is the first step. Recent advances in knowledge of local bottom communities make it desirable to revise community concepts developed a few years ago, because: (1) attached vegetation, which was the basis for community zonation originally, has practically disappeared from the areas of chief interest; (2) the two plant-animal communities originally described can now be combined and redescribed as one hard-bottom community of animals with its clean-water parallel (*Laevicardium-Codakia* community) a few miles south in Biscayne Bay; (3) the *Ophionephthys limicola* zone originally discerned is in fact a derivative of the clean-water *Amphioplus-Ophionephthys* community, also a few miles south (McNulty, 1961; McNulty *et al.,* 1962a).

The community approach requires knowledge of the preferential selection of sediments by the biota. Such knowledge is particularly important in studies of pollution because of the necessity of separating effects of pollution from effects of substratum. For example, ecological observations on the sewage indicator *Capitella capitata* show that this polychaete is found strictly in very fine sediments and that it does not tolerate any fraction of coarse material (Bellan, 1964). Another example is the difficulty encountered by Alexander *et al.* (1936) in distinguishing biotic changes due to pollution from those due to differences in the substratum.

The total of 137 station samples taken in both pre- and postabatement samplings in this study, together with data for particle sizes at each station, provide excellent material for investigation of animal-sediment relationships. Some species can be arranged in a selectivity spectrum that covers a range from fine-sediment selective to coarse-sediment selective, while other species seem to be essentially nonselective (Figs. 4 and 5). All stations were arranged in order according to their sediments, from fine to coarse, by median grain diameter. They were then divided into three groups of equal size: the relatively fine-sediment group (0.05 to 0.30 mm median grain diameter), the medium-sediment group (0.31 to 0.47 mm), and the relatively coarse-sediment group (0.48 to 0.86 mm). A tally

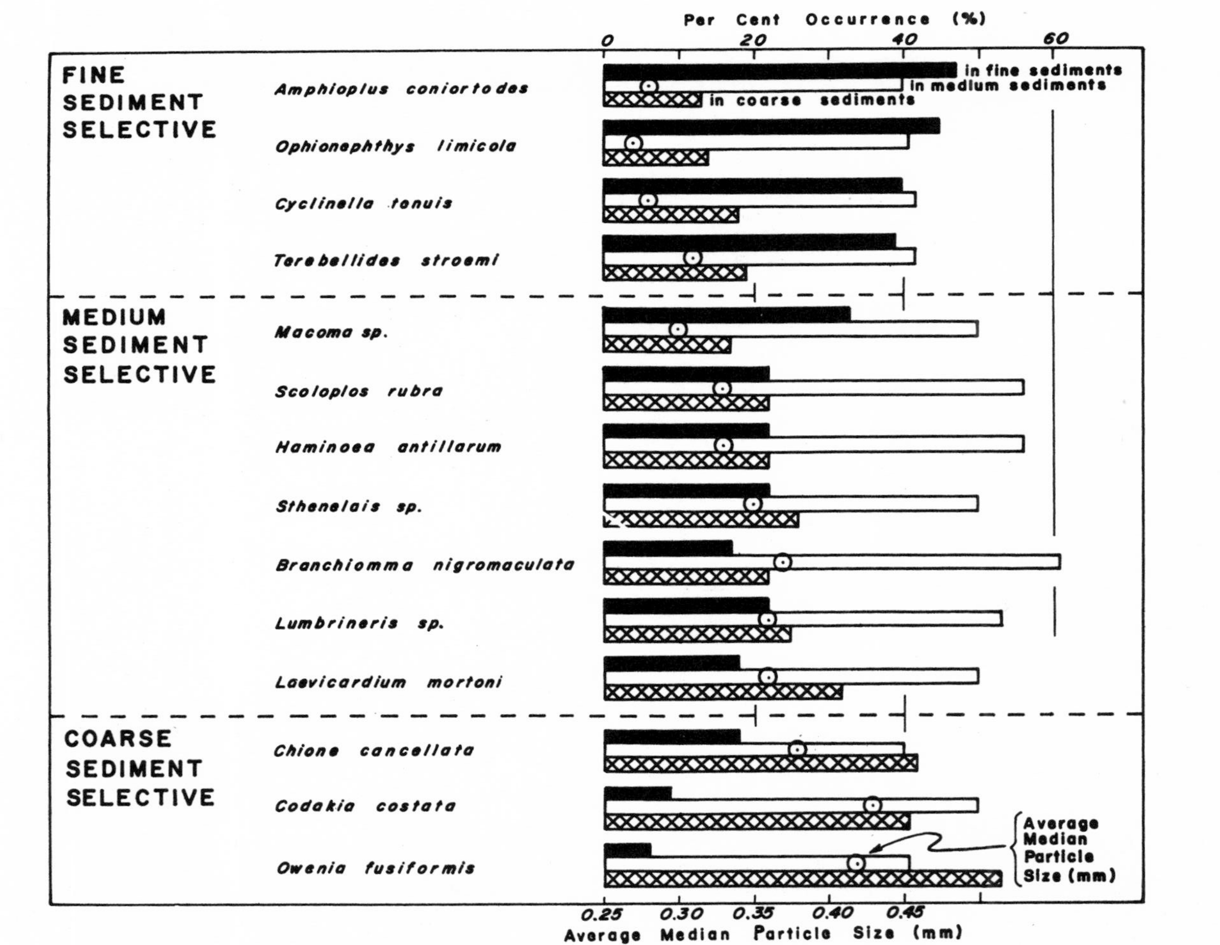

Fig. 4. Sediment-selectivity spectrum of species that select fine, medium, and coarse sediment

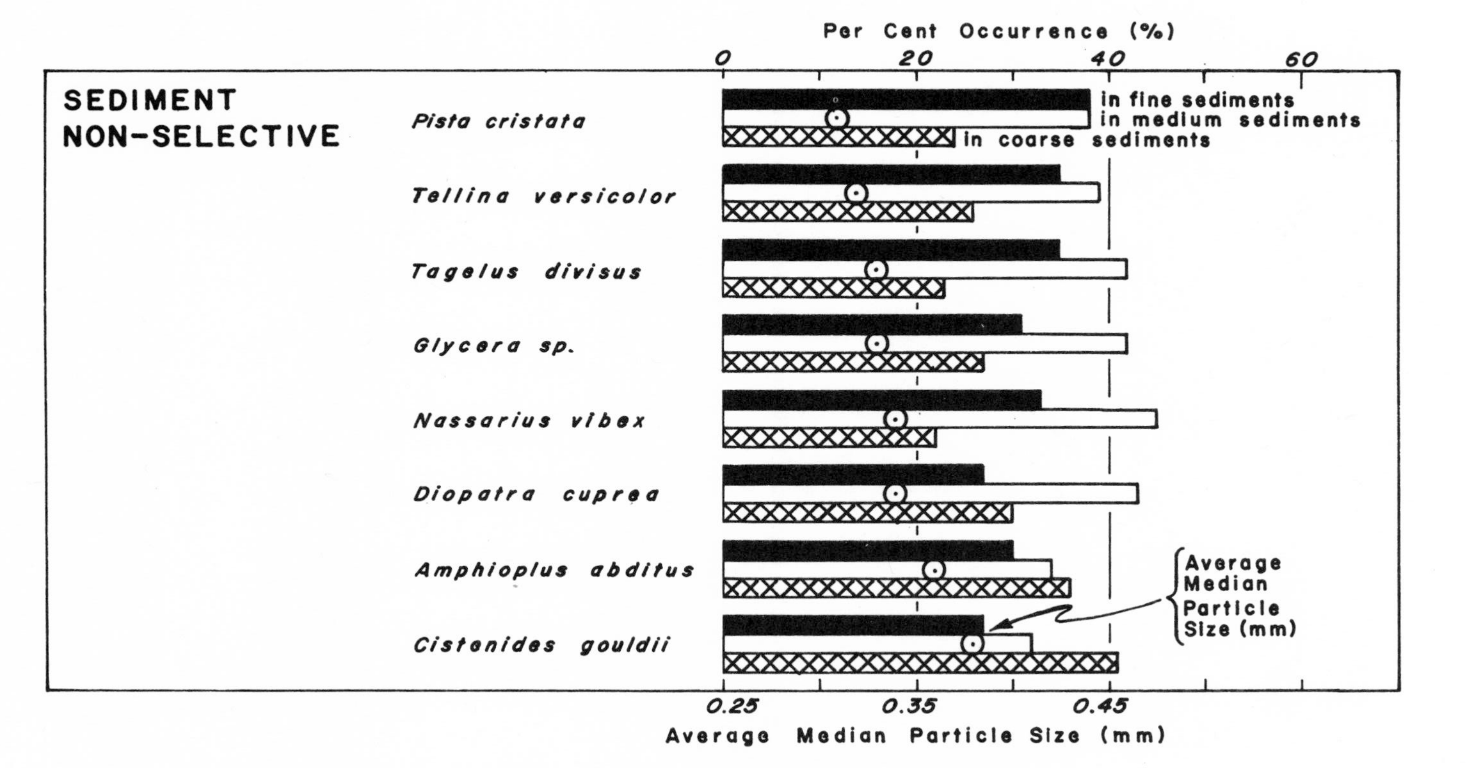

Fig. 5. Sediment-selectivity spectrum of essentially nonselective species

of the occurrence of the various species produced the arrays shown, which indicate that type of sediment strongly affects the occurrence of some species but only weakly affects the occurrence of other species. This information was then used in the community descriptions that follow.

Attention has been centered upon the 31 stations in, and adjacent to, the formerly most heavily polluted parts of the bay. Nine stations with median grain diameters of 0.05 to 0.22 mm were designated "soft-bottom stations." The remaining 22 stations with median grain diameters of 0.28 to 0.83 mm were designated "hard-bottom stations"; this group included a few stations transitional between truly soft-bottom and truly hard-bottom stations, but included only two stations with medians greater than 0.50 mm. Tables 3 and 4 provide arithmetic mean numbers of individuals per square meter for both polluted (1956) and unpolluted (1960) conditions at various distances from former outfalls. Figure 6 shows the locations of the ecological zones labeled in the tables "X" and "Y" (soft-bottom areas), and "A" through "E" (hard-bottom areas).

TABLE 3

Comparison of Macrobenthos of Soft-Bottom Areas (median particle-size 0.05 - 0.22 mm) in Northern Biscayne Bay, Florida, for July-August, 1956 (polluted) and November-December, 1960 (unpolluted) in Zones X and Y (See Figure 6)

	Zone X 50–100 m from outfalls; Stas. 15, 17, 18		Zone Y 370–1,850 m from outfalls; Stas. 403W, 403, 404, 502 503, 603	
Species	1956 n/m^2	1960 n/m^2	1956 n/m^2	1960 n/m^2
POLYCHAETA				
Branchiomma nigromaculata	–	–	9.0	–
Glycera sp.	–	1.5	3.8	2.3
Lumbrineris sp.	–	1.5	–	–
Lysidice ninetta	–	1.5	–	–
Pista cristata	1.5	–	6.0	–
Polydora sp.	1.5	–	2.3	–

Table 3, continued

Species	Zone X 1956 n/m²	Zone X 1960 n/m²	Zone Y 1956 n/m²	Zone Y 1960 n/m²
Sthenelais sp.	–	–	0.8	–
Terebellides stroemi	–	–	8.2	–
Scoloplos rubra	–	–	–	2.3
Chaetopterus variopedatus	–	–	0.8	–
Unidentified polychaetes	–	1.5	–	–
SIPUNCULIDA				
Unidentified sipunculid	–	–	0.8	–
AMPHIPODA				
Grubia filosa	–	–	(7.5)[1]	–
Grubia sp.	–	–	(1.6)[1]	–
BRACHYURA				
Grapsoid crab	–	–	1.6	–
GASTROPODA				
Bulla occidentalis	–	–	1.6	–
Nassarius vibex	–	–	3.1	–
Olivella perplexa	–	–	–	0.8
LAMELLIBRANCHIATA				
Chione cancellata	–	–	7.5	1.5
Cyclinella tenuis	–	–	–	5.2
Dosinia elegans	–	–	–	1.5
Macoma tenta	–	–	–	3.1
Nucula proxima	–	–	–	6.0
Tagelus divisus	4.5	4.5	4.6	29.2
Tellina alternata	–	–	4.6	–
T. martinicensis	–	–	–	2.3
T. versicolor	1.5	–	24.0	13.0
OPHIUROIDEA				
Amphioplus abditus	–	–	0.8	4.4
A. coniortodes	–	–	0.8	–
Amphipholis gracillima	–	–	–	2.3
Ophionephthys limicola	–	–	6.7	7.5

Table 3, continued

Species	Zone X 1956 n/m²	Zone X 1960 n/m²	Zone Y 1956 n/m²	Zone Y 1960 n/m²
ASCIDIACEA				
Unidentified ascidian	–	–	1.6	–
Total numbers per square meter[1]	9.0	10.5		81.4

[1]Amphipods excluded from total.

TABLE 4

Comparison of Macrobenthos of Hard-Bottom Areas (median particle-size 0.28 - 0.83 mm) in Northern Biscayne Bay, Florida, for July-August, 1956 (polluted) and November-December, 1960 (unpolluted) in Zones A through E (See Figure 6)

Species	Zone A 20 m from outfalls		Zone B 100-370 m from outfalls		Zone C 185-740 m from outfalls		Zone D 925-2,220 m from outfalls		Zone E 1,300-4,100 m from outfalls	
	Sta. 401W		Stas. 703, 704, 704E		Stas. 401, 402, 402E, 501, 16, 706		Stas. 604, 605, 606, 707, 20		Stas. 607, 608, 609, 708, 20E, 73W, 73	
	1956 n/m²	1960 n/m²	1956 n/m²	1960 n/m²	1956 n/m²	1960 n/m²	1956 n/m²	1960 n/m²	1956 n/m²	1960 n/m²
POLYCHAETA										
Branchiomma nigromaculata	–	–	–	–	34.5	–	9.1	–	1.3	–
Cistenides gouldii	–	22.5	1.5	–	3.8	2.3	–	0.9	–	2.6
Diopatra cuprea	–	–	24.2	3.0	7.6	1.6	–	–	–	1.3
Glycera sp.	–	–	6.0	4.5	35.2	11.2	7.3	2.7	1.9	1.9
Lumbrineris sp.	–	–	–	1.5	2.3	2.3	3.7	6.2	1.9	1.9
Maldane sarsi	–	–	–	–	20.2	–	–	–	1.9	–
Naineris setosa	4.5	–	45.0	–	21.0	–	–	–	–	–
Onuphis magna	–	–	3.0	–	4.6	–	–	–	3.2	–

Table 4, continued

Species	[illegible] n/m^2	1960 n/m^2	1956 n/m^2	1960 n/m^2	1956 n/m^2	1960 n/m^2	1956 n/m^2	1960 n/m^2	1956 n/m^2	1960 n/m^2
Owenia fusiformis	4.5	9.0	6.0	–	10.5	5.2	9.2	5.5	5.1	11.6
ista cristata	-	–	–	–	3.8	–	–	–	0.6	–
Polydora s	–	–	45.0	–	17.2	–	–	–	–	–
Semiodera	–	–	1.5	–	0.8	–	–	–	–	0.6
Sthenelais		–	7.5	–	12.0	–	2.7	–	3.9	–
Scoloplos	–	–	–	1.5	–	2.3	–	0.9	–	1.3
Terebelli	–	–	–	–	9.7	–	3.7	–	12.2	1.3
Unidentifi	–	4.5	10.5	6.0	30.0	2.4	4.5	3.7	7.1	1.9
SIPUNCUL										
Golfingia sp.	–	18.0	–	–	–	0.8	–	–	–	–
Phascolion sp.	–	4.5	–	–	–	–	–	–	–	–
Unidentified si[illegible]d	–	–	–	–	5.3	2.3	–	–	–	1.3
NEMERTEA										
Unidentified nemertean	13.5	–	–	–	–	–	–	–	–	–
AMPHIPODA										
Corophium acherusicum	–	–	–	–	(86.2)*	–	–	–	–	–
Grubia filosa	–	–	–	–	(1.6)*	–	(3.7)*	(1.3)*	–	–
Grubia sp.	–	(1.5)*	–	–	(1.6)*	–	–	–	–	–
TANAIDACEA										
Unidentified tanaidacean	–	(75.0)*	–	–	(71.9)*	–	(2.7)*	–	–	–
ISOPODA										
Unidentified isopod	–	–	–	–	(3.1)*	–	–	–	–	–
ANOMURA										
Hermit crab	–	1.5	–	–	–	–	0.9	–	–	–
BRACHYURA										
Eucratopsis crassimanus	–	–	3.0	–	–	0.8	–	0.9	1.9	–
Grapsoid crab	–	–	–	–	3.8	–	–	–	–	–
Spider crab	–	–	–	–	3.8	–	1.8	–	–	–
PENAEIDEA										
Penaeus duorarum	–	4.5	–	–	–	–	–	–	–	–
Trachypenaeus constrictus	–	–	–	–	–	0.8	–	–	–	–
CARIDEA										
Alpheus floridanus	–	–	–	–	–	0.8	–	3.6	–	0.6
Snapping shrimp	–	–	–	–	–	–	1.8	–	4.5	–
Unidentified caridean	–	–	–	–	1.5	–	–	–	–	1.2
GASTROPODA										
Anachis avara	–	–	–	–	2.2	–	–	–	–	–
Bulla occidentalis	–	–	–	–	7.5	–	–	–	0.6	–
B. striata	–	–	–	–	–	1.8	–	–	–	1.2

Table 4, continued

Species	1956 n/m2	1960 n/m2	1956 n/m2	1960 n/m2	1956 n/m2	1960 n/m2	1956 n/m2	1960 n/m2	1956 n/m2	1960 n/m2
Conus stearnsii	–	–	–	–	–	–	–	–	1 2	–
Epitonium rupicolum	4.5	–	–	–	–	–	–	–	–	–
Haminoea antillarum	–	–	1.5	–	10.5	–	–	–	–	–
Nassarius vibex	4.5	–	76.5	–	25.4	–	0.9	–	–	–
Neritina virginea	–	–	16.5	–	–	–	–	–	–	–
LAMELLIBRANCHIATA										
Botula castanea	–	–	–	–	3.8	–	–	–	–	–
Cardita floridana	–	4.5	–	–	3.1	–	–	–	0.6	–
Chione cancellata	–	–	25.7	12.0	169.2	9.7	25.4	7.3	11.7	11.0
Codakia costata	–	–	–	–	–	36.7	–	40.0	–	21.2
Cumingia tellinoides	–	–	–	–	–	–	1.6	–	–	–
Cyclinella tenuis	–	–	–	1.5	–	2.3	–	2.7	–	4.5
Dosinia elegans	–	–	–	–	4.6	0.8	–	0.9	–	–
Laevicardium mortoni	–	–	1.5	–	7.6	1.6	3.7	1.8	1.9	2.6
Lioberus castaneus	–	–	–	–	–	1.6	–	–	–	–
Lima pellucida	–	–	–	–	3.1	–	–	–	–	–
Lucina multilineata	–	–	–	–	8.9	–	–	–	–	–
Lucina sp.	–	–	–	–	–	–	–	–	–	0.6
Macoma tenta	–	–	–	1.5	–	–	–	–	–	1.2
Macoma sp.	–	–	1.5	–	3.1	–	2.3	–	–	–
Mactra fragilis	–	–	–	–	2.3	–	–	–	–	–
Modiolus papyria	–	–	–	–	0.8	–	–	–	–	–
Nucula proxima	–	–	–	–	–	0.8	–	4.5	–	–
Tagelus divisus	–	40.5	28.5	3.0	51.6	12.0	32.8	13.6	6.4	7.7
Tellina alternata	–	–	–	–	–	–	2.3	–	4.5	–
T. lineata	–	–	–	–	–	–	–	–	–	0.6
T. martinicensis	–	–	–	–	–	–	–	–	–	0.6
T. versicolor	–	18.0	37.5	1.5	95.2	0.8	17.3	3.7	9.7	1.9
Tellina sp.	–	–	9.0	–	0.8	–	0.9	–	–	–
Trachycardium egmontianum	–	–	1.5	–	0.8	–	–	–	–	–
OPHIUROIDEA										
Amphiodia pulchella	–	–	–	–	–	–	–	0.9	–	–
Amphioplus abditus	–	–	–	–	1.6	0.8	2.3	1.8	3.9	3.2
A. coniortodes	–	–	4.5	–	–	–	0.9	–	1.3	–
Amphipholis gracillima	–	–	–	–	–	–	–	1.8	–	–
Ophionephthys limicola	–	–	–	1.5	0.8	5.9	0.9	3.7	0.9	2.6
ASCIDIACEA										
Unidentified ascidian	–	–	–	–	20.2	–	1.8	–	3.9	–
TOTALS*	18.0	126.0	358.9	37.5	650.7	107.6	137.8	107.1	92.1	86.4

*Amphipods, tanaidaceans, and isopods excluded from totals.

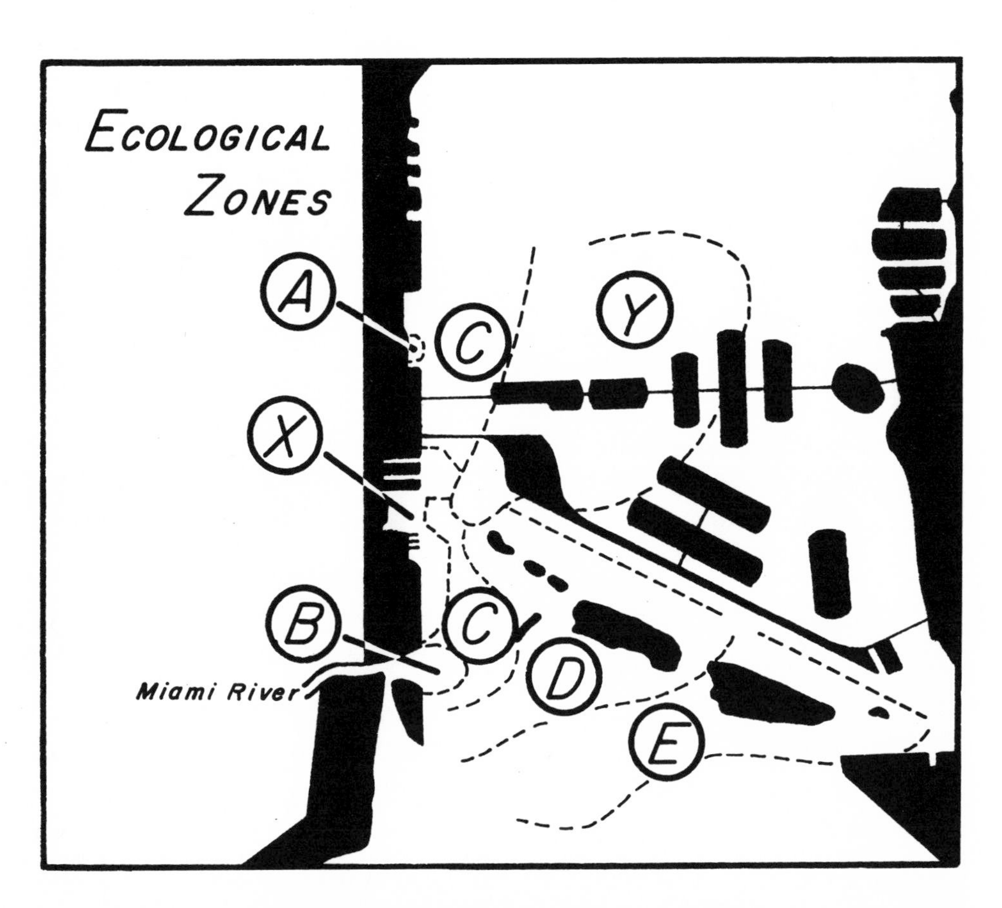

Fig. 6. Ecological zones. X and Y—soft bottom;
A through E—hard bottom

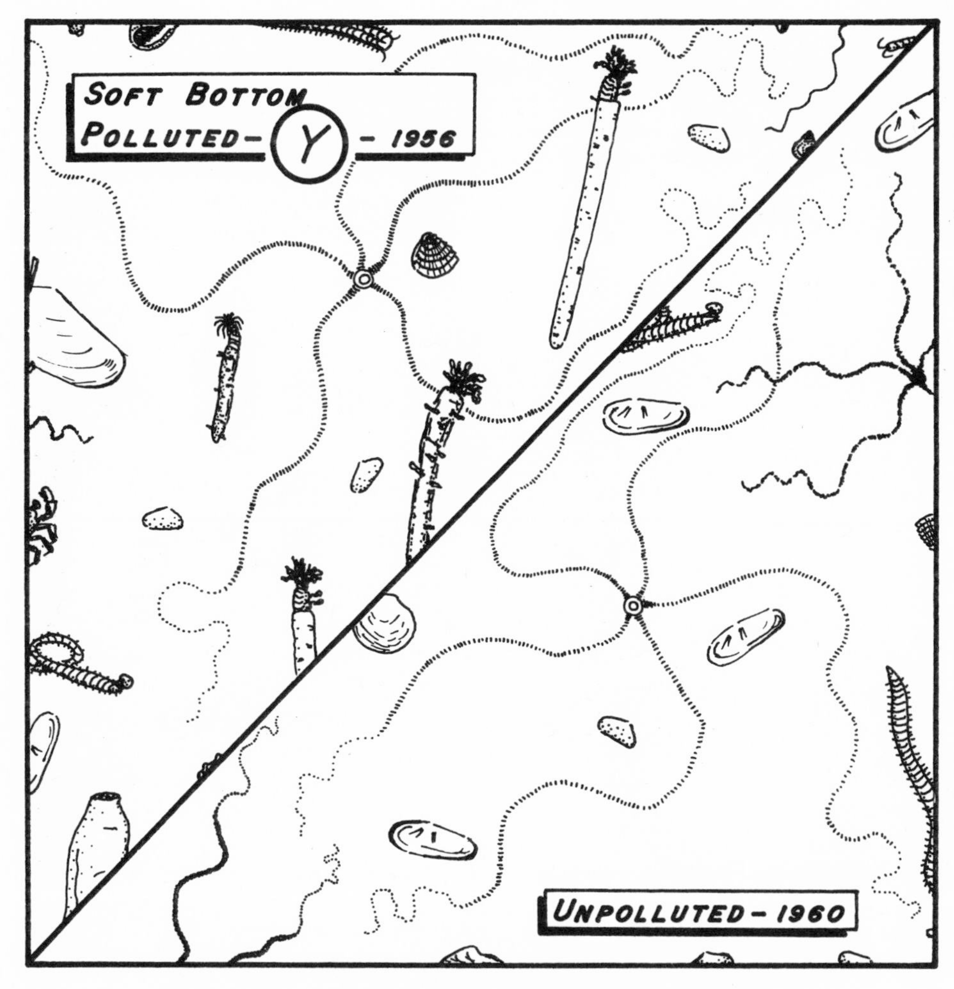

Fig. 7. Pre- and postabatement populations in 1/4 m^2 of soft bottom 370 to 1,850 m from outfalls, Zone Y

Figures 7, 8, 10, 11, and 12 compare benthic fauna under both polluted and unpolluted conditions in typical 1/4-m^2 plots. Fractions of animals in the figures represent calculated fractions of animals for the areas shown. Ecological zones referred to by circled letter designations are shown in Figure 6.

The Soft-Bottom Community

The structure of the normal soft-bottom community is assumed to be represented by the assemblage of Zone Y found in 1960 (Fig. 7). It is a derivative of the *Amphioplus-Ophionephthys* community of soft-bottom areas a few miles south in the bay, differing from it chiefly in the scarcity or absence of *Amphioplus coniortodes.* Community description, based on criteria suggested by Thorson (1957: 477), is as follows:

First-order species:	*Ophionephthys limicola*
Second-order species:	*Cyclinella tenuis, Nucula proxima*
Third-order species:	*Glycera* sp., *Tagelus divisus, Tellina versicolor*
Associated species:	*Amphioplus abditus, Amphipholis gracillima, Macoma tenta, Tellina martinicensis*

With moderate pollution, major changes in the community were a marked increase in the number of immigrant, noncommunity species *(Branchiomma nigromaculata, Chione cancellata, Terebellides stroemi, Grubia filosa, Grubia* sp., *Nassarius vibex, Pista cristata, Polydora* sp., and *Sthenelais* sp.), plus the absence of certain community species (*A. gracillima, C. tenuis, M. tenta, N. proxima,* and *T. martinicensis*). There were 21 species with pollution, and 14 species without; 88.6 individuals per m^2 with pollution, and 81.4 without.

An extremely impoverished fauna typified the soft-bottom area next to the bay's western shore, practically in the shadow of former outfalls (Fig. 8). Here the normal soft-bottom community was not found, even without pollution. With pollution, three species were taken *(Pista cristata, Polydora* sp., and *Tellina versicolor)* that were not taken later in clean-water sampling, when the species sampled were *Glycera* sp., *Lumbrineris* sp., *Lysidice ninetta,* and *Tagelus divisus. T. divisus* was taken in both samplings. Net changes were small: 4 species with pollution, and 4+ species without; 9.0 individuals per m^2 with pollution, and 10.5+ without.

Figure 9 is a diagrammatic representation of average total numbers of individuals in soft-bottom areas plotted against distance bayward from former outfalls. It demonstrates almost identical average numbers of individuals per m^2 under both polluted and unpolluted conditions.

The author's general conclusion on soft-bottom areas is that changes were surprisingly small. Moderate pollution was accompanied by an influx of various

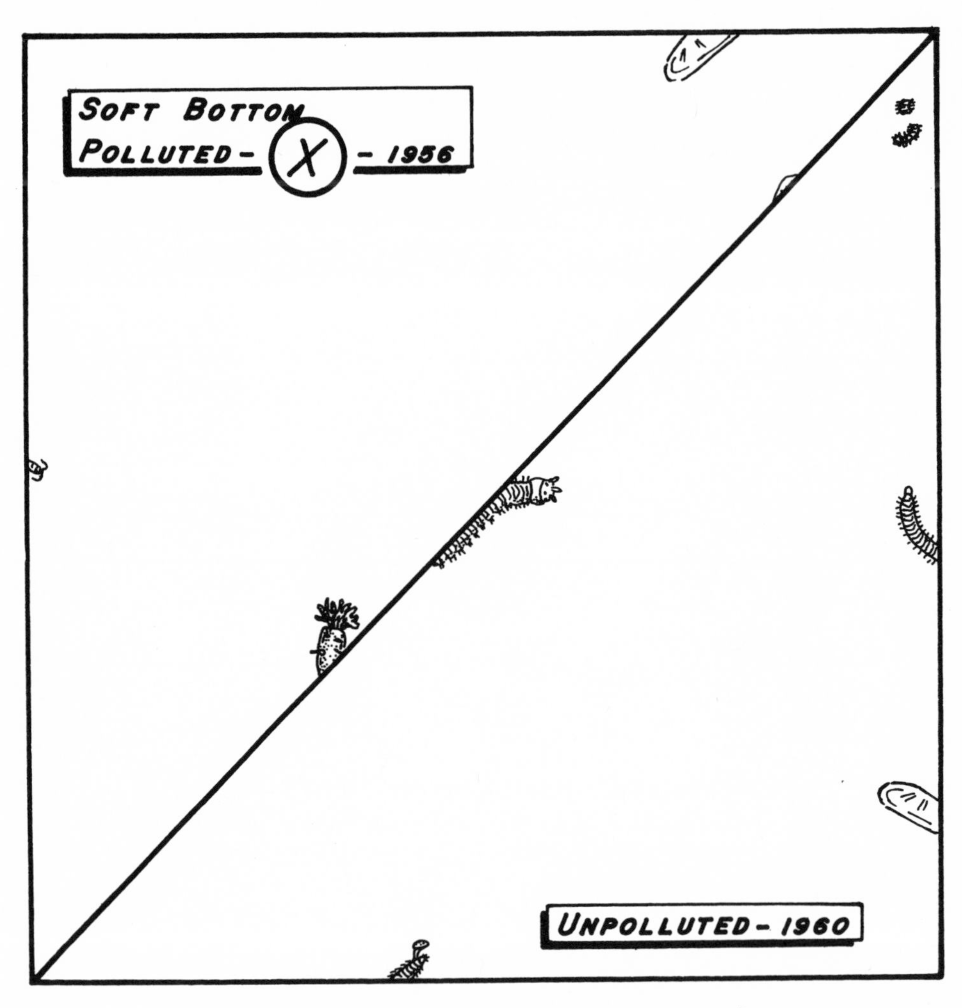

Fig. 8. Pre- and postabatement populations in 1/4 m^2 of soft bottom 50 to 100 m from outfalls, Zone X

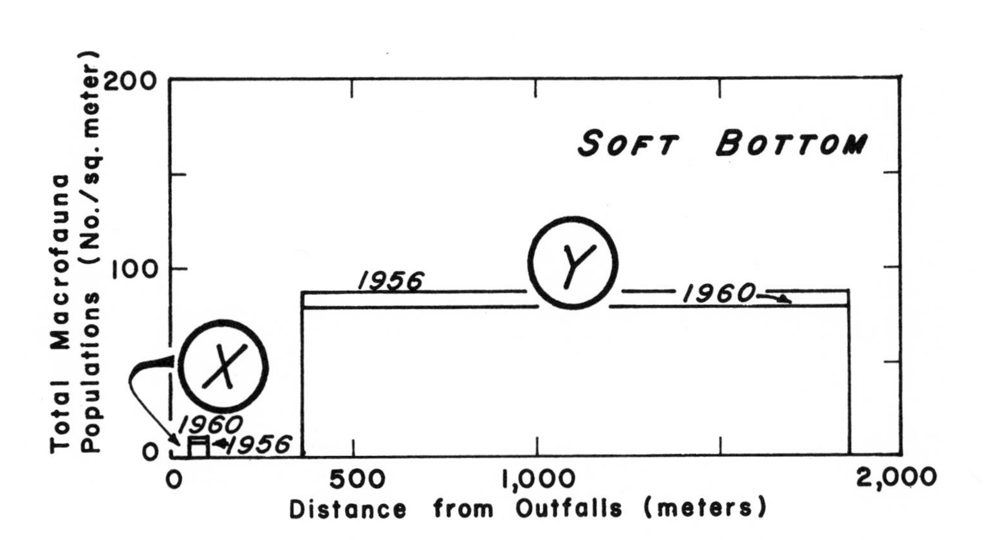

Fig. 9. Diagram of mean total numbers of individuals per m^2 in soft-bottom areas plotted against distance seaward from outfalls, zones X and Y, for the years 1956 and 1960

species of polychaetes (mainly) and a disappearance of various molluscan species at distances of 370 to 1,850 m from outfalls. The amphiuran fauna, present under both polluted and unpolluted conditions, was impoverished compared with soft-bottom areas southward in the bay. In deeper areas closer to outfalls (50 to 100 m), the mean number of individuals per m^2 remained almost identical both with and without pollution, although there was a complete shift in species of polychaetes. Here, fresh groundwater entering the bay along an orifice and through springs along the bay's western shore may permanently depress infaunal populations (Kohout & Kolipinski, personal communication).

The Hard-Bottom Community

The structure of the normal hard-bottom community is assumed to be represented by the assemblage of Zone E, found in 1960 (Fig. 10), as follows:

First-order species:	*Chione cancellata*
Second-order species:	*Laevicardium mortoni* and *Codakia costata*
Third-order species:	*Owenia fusiformis, Tagelus divisus,* and *Tellina versicolor*
Associated species:	*Amphioplus abditus, Cistenides gouldii, Glycera americana,* and *Lumbrineris maculata*

With pollution, populations were either greatly increased or decreased, depending on proximity of outfalls and hydrography. In Zone C, which lies some 185 to 740 meters bayward of former outfalls and across which vigorous tidal currents up to about 0.6 m/sec typically sweep, there was a pronounced fertilizing effect with pollution (Fig. 11). With pollution, all community species except *Codakia costata* were present, and to the community species were added many immigrant species of polychaetes, tanaidaceans, amphipods, lamellibranchs, gastropods, and ascidians. The immigrants included mainly *Branchiomma nigromaculata, Diopatra cuprea, Maldane sarsi, Naineris setosa, Onuphis magna, Pista cristata, Polydora* sp., *Semiodera roberti, Sthenelais* sp., *Terebellides stroemi, Corophium acherusicum, Grubia filosa, Grubia* sp., *Dosinia elegans, Lucina multilineata, Bulla occidentalis, Haminoea antillarum,* and *Nassarius vibex,* plus unidentified ascidians and tanaidaceans. The number of species with pollution was at least 46 compared with at least 25 species without pollution; numbers of individuals per m^2 were 651 and 108 for polluted and unpolluted samplings, respectively, excluding amphipods and tanaidaceans.

The reverse of the fertilizing effect described above was observed in Zone A, within 20 m of an outfall (Fig. 12). Without pollution, 4 of the 10 community species were taken: *Cistenides gouldii, Owenia fusiformis, Tagelus divisus,* and

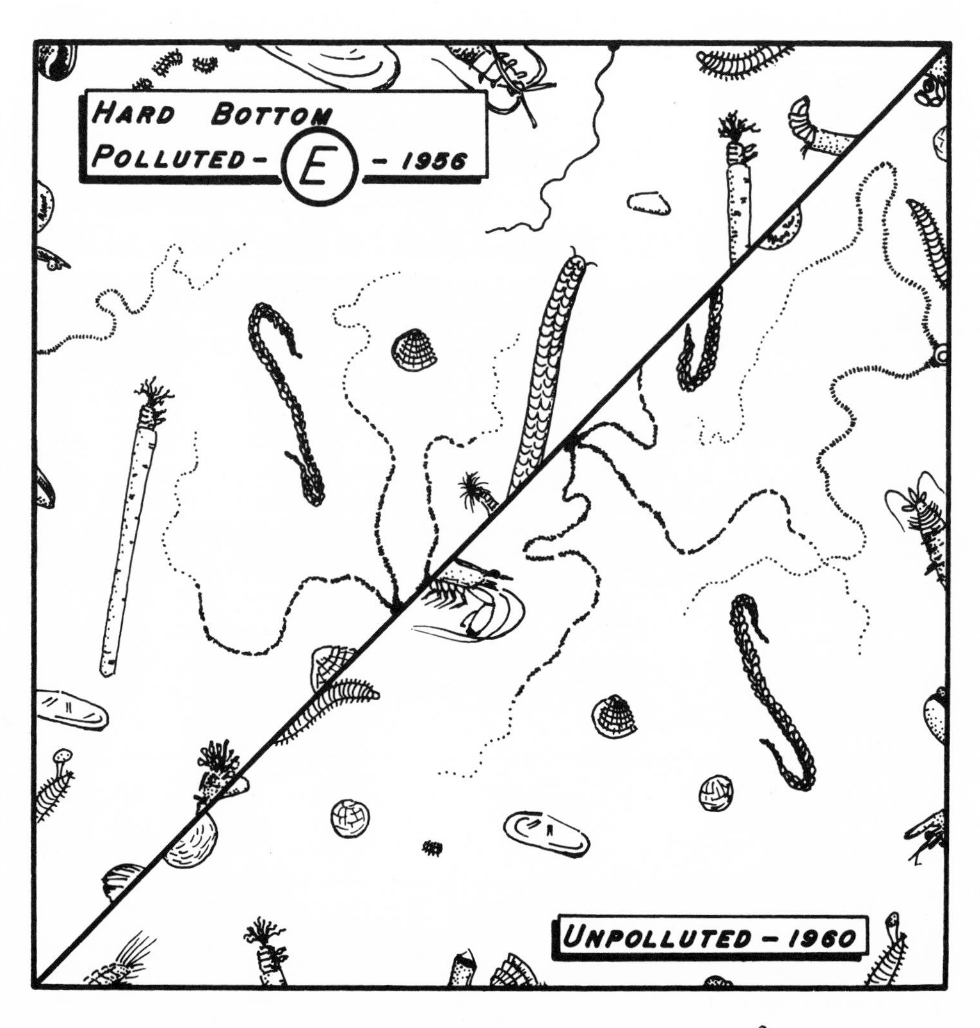

Fig. 10. Pre- and postabatement populations in 1/4 m^2 of hard bottom 1,300 to 4,000 m from outfalls, Zone E

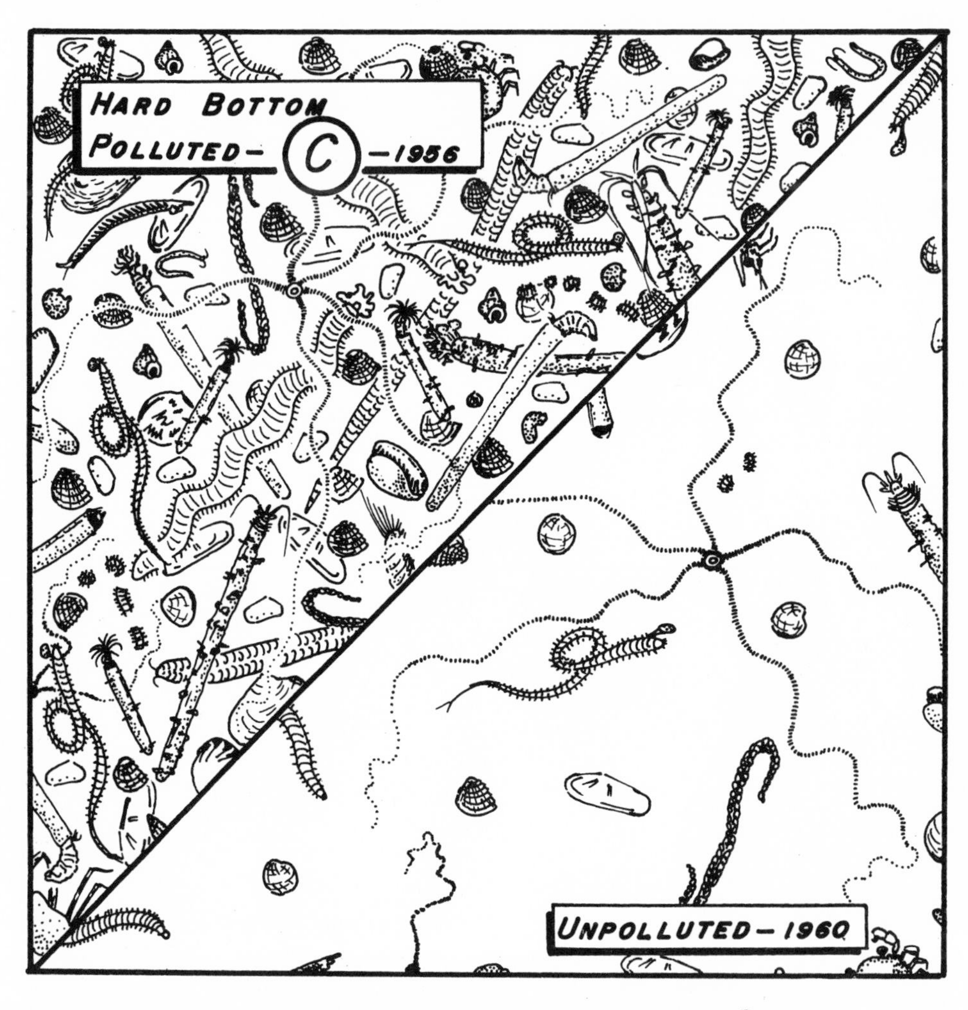

Fig. 11. Pre- and postabatement populations in 1/4 m^2 of hard bottom 185 to 740 m from outfalls, Zone C

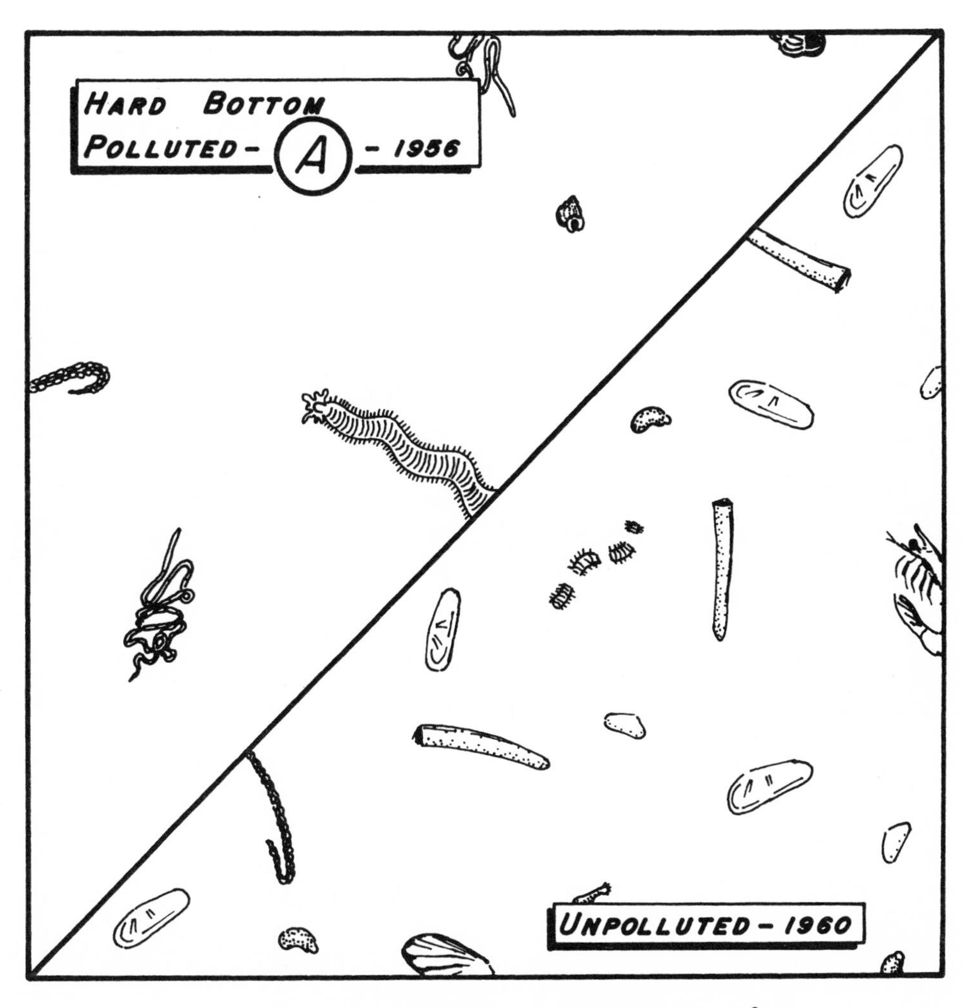

Fig. 12. Pre- and postabatement populations in 1/4 m^2 of hard bottom 20 m from an outfall, Zone A

T. versicolor. With pollution, only *O. fusiformis* of the community species was taken, together with the immigrant species *Epitonium rupicolum, Naineris setosa, Nassarius vibex,* and an unidentified nemertean. In summary, in Zone A there were at least 5 species with, and at least 9 species without pollution; there were approximately 18 individuals per m^2 with pollution and approximately 126 without. The aberrant community structure found in this zone under clean water conditions could result from upwelling of fresh groundwater through the bottom near shore, as considered likely for soft-bottom areas (above) within roughly 122 m of shore.

The normal community described above is somewhat like the hard-bottom assemblage named the *Laevicardium-Codakia* community to the south in Biscayne Bay, and differs from it most notably by having more species and higher population density.

TABLE 5

Comparison of Volumes of Major Groups of Benthic Plants in Hard-Bottom Areas (median particle-size 0.05 - 0.22 mm) in Northern Biscayne Bay, Florida, for July-August, 1956 (polluted) and November-December, 1960 (unpolluted), in Zones A through E (See Figure 6)

	Zone A 20 m from outfalls		Zone B 100-370m from outfalls		Zone C 185-740m from outfalls		Zone D 925-2,220m from outfalls		Zone E 1,300-4,100m from outfalls	
	Sta. 401W		Stas. 703, 704, 704E 501, 16		Stas. 401, 402, 402E, 501, 16, 706		Stas. 604, 605, 606, 707, 20		Stas. 607, 608, 609, 708, 20E, 73W, 73	
Plant Group	1956 ml/m^2	1960 ml/m^2	1956 ml/m^2	1960 ml/m^2	1956 ml/m^2	1960 ml/m^2	1956 ml/m^2	1960 ml/m^2	1956 ml/m^2	1960 ml/m^2
Spermatophytes	–	–	–	–	81.8 *	–	100.5	–	292.0	106.1
Red algae	trace	–	12.0	–	60.8	–	0.9	–	18.0	3.2
Other	–	–	4.5	–	15.0	–	–	–	9.1	–
TOTALS	–	–	16.5	–	157.6	–	101.4	–	319.1	109.3

*Displacement volume in milliliters.

Figure 13 is a diagrammatic representation of the average total numbers of individuals per m^2 in Zones A through E plotted against distance bayward from outfalls. It demonstrates the depressant effect of pollution near outfalls and its fertilizing effect farther away, until the latter effect becomes indetectable.

Attached Vegetation

Decline of attached vegetation in formerly polluted areas of hard bottom is shown in Table 5, a summary of calculated volumes of benthic plant material per m^2 for both 1956 and 1960. (Soft-bottom areas were essentially devoid of attached vegetation in both samplings.)

The data for spermatophytes alone are plotted in Figure 14 to provide a diagrammatic display of the relative abundance of average seagrass volumes at varying distances bayward from former outfalls. The species are *Diplanthera wrightii, Halophila baillonis,* and *Thalassia testudinum.*

One would expect that, with elimination of pollution, conditions should favor increased growth of algae and seagrasses. The opposite actually occurred between 1956 and 1960, however, probably due mainly to increased turbidity and sedimentation from hydraulic dredging for construction of the Julia Tuttle Causeway.

Inorganic Phosphate-Phosphorus

In the preabatement study, a high content of dissolved inorganic phosphate-phosphorus was clearly associated with pollution. The highest values were found in the Miami River water, which was also the most polluted water, and along the northwestern shoreline, which was highly polluted and poorly flushed by tidal action (McNulty, 1960). Postabatement concentrations were much less, as anticipated. Concentrations a little less than half those formerly encountered were found in Miami River water, which is rich with detritus from land runoff and which still receives some sewage from private outfalls. The latter is negligible in quantity compared with preabatement quantities from public outfalls, as previously mentioned. Dilution is rapid, however, and concentrations quickly drop in the seaward direction, as shown in Table 6 and Figure 15.

Volumes of Zooplankton

Volumes of zooplankton have decreased most markedly in the north bay areas where formerly they exceeded by a factor of 10 the volumes encountered in the most southerly, unpolluted parts of the study area. The decrease is greatest

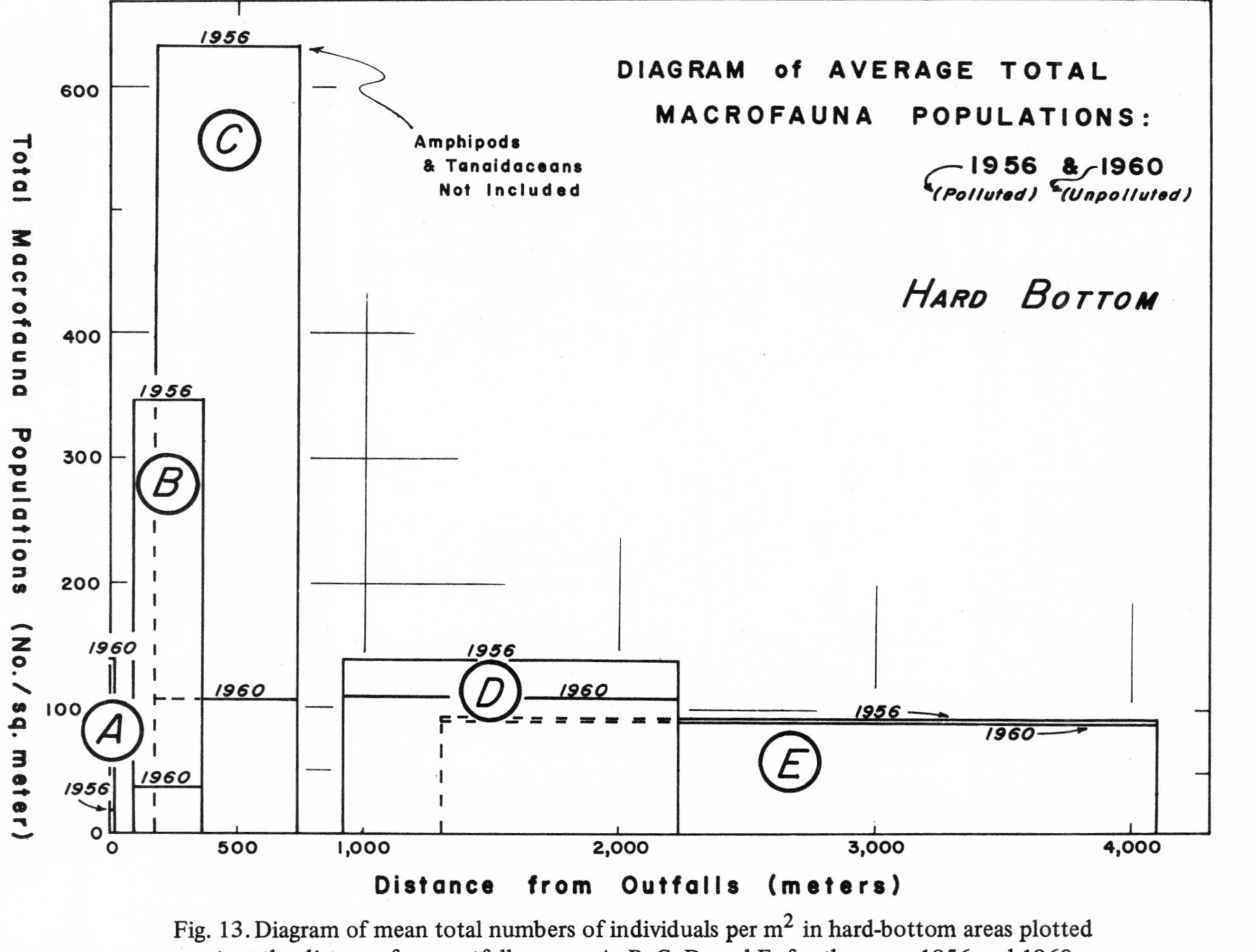

Fig. 13. Diagram of mean total numbers of individuals per m^2 in hard-bottom areas plotted against the distance from outfalls, zones A, B, C, D, and E, for the years 1956 and 1960

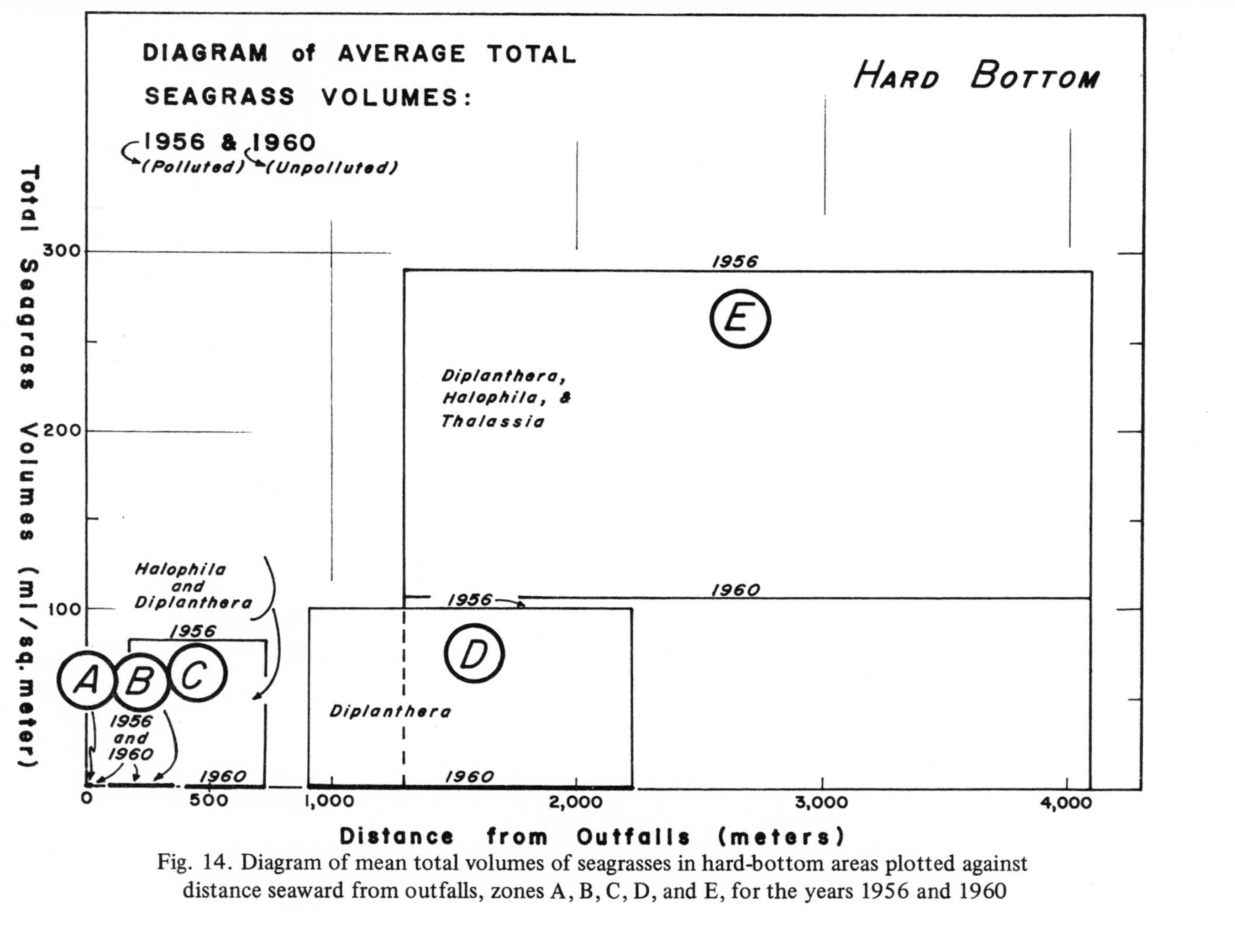

Fig. 14. Diagram of mean total volumes of seagrasses in hard-bottom areas plotted against distance seaward from outfalls, zones A, B, C, D, and E, for the years 1956 and 1960

near the northwesterly shore, where phosphate concentrations likewise have decreased markedly. It is especially interesting that volumes have remained nearly the same over much of the study area, indicating that pollution-related factors stimulated production of zooplankton only in the poorly flushed, extremely shallow, north bay area (Table 7, Fig. 16).

TABLE 6

Mean Dissolved Inorganic Phosphate-Phosphorus

Sta.	(1) Postabatement mean PO_4-P (μg-a/l)†	(2) Number of observations	(3)* Preabatement mean PO_4-P (μg-a/l)	(4) Difference (1)–(3) (μg-a/l)
801	0.18	2	0.37	–0.19
703	0.49	5	1.07	–0.58
20	0.22	5	0.53	–0.31
708	0.18	6	0.23	–0.05
710	0.23	5	0.14	+0.09
608	0.14	6	0.15	–0.01
605	0.17	6	0.41	–0.24
16	0.35	6	0.50	–0.15
401	0.28	6	0.62	–0.34
304	0.18	6	0.40	–0.22
8	0.23	6	1.10	–0.87
88	0.23	6	0.17	+0.06
406	0.24	6	0.16	+0.08

*From McNulty (1960).

†To convert from microgram-atoms per liter to parts per million, multiply by 0.031.

Fouling Organisms

Tables 8, 9, and 10 summarize the data on fouling organisms. Mean volumes of fouling organisms (Table 8 and Fig. 17) indicate both decreases and increases, hence no clear relationship to pollution. Mean abundance of barnacles (Table 9 and Fig. 18) likewise indicates no clear relationship to pollution. The mean numbers of amphipod tubes (Table 10 and Fig. 19), however, appear to show that

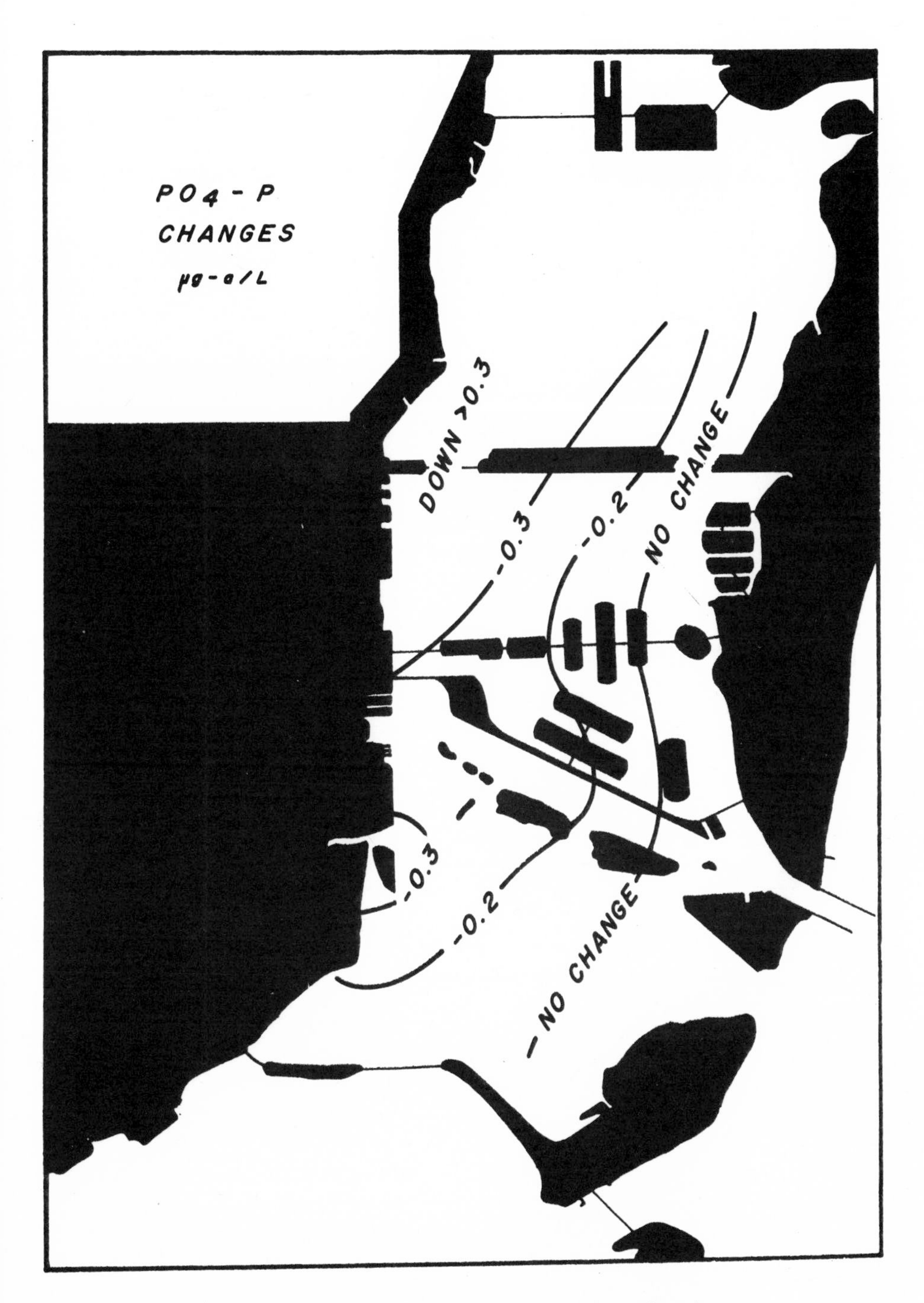

Fig. 15. Changes in mean dissolved inorganic phosphate-phosphorus; minus sign indicates decrease after abatement of pollution

amphipod tubes were more abundant under polluted than under unpolluted conditions.

In general, the results follow a pattern seen in past studies here and elsewhere. Abundance of barnacles over many years in northern Biscayne Bay has followed an upward trend that appears to be related to the availability of surfaces of attachment, such as pilings and sea walls (Moore & Frue, 1959). In past pollution studies, the availability of such surfaces has clearly been a controlling factor in the abundance of barnacles (McNulty, 1957, 1961; Smith *et al.*, 1950). Amphipods, on the other hand, have consistently been more abundant in polluted than in unpolluted parts of the north bay area (McNulty, 1957, 1961). The same results have been noted by Barnard (1958), Barnard & Reish (1959), and Waldichuk & Bousfield (1962) in coastal areas of the western states. The author has encountered huge concentrations of benthic amphipods blanketing the bottom of McKay Bay, an arm of Tampa Bay polluted heavily with organic wastes from meat rendering plants and domestic sewage. The species of amphipods known to occur on the Biscayne Bay panels are *Erichthonius brasiliensis* (Dana), *Corophium acherusicum* (Costa), *Elasmopus pectenicrus* (Bate), *Podocerus brasiliensis* (Dana), and *Grubia filosa* (Savigny) (see McNulty, 1961).

TABLE 7

Displacement Volumes of Zooplankton Using Clarke-Bumpus Sampler with Number Two Net (aperture 0.366 mm)

Sta.	(1) Postabatement mean volume (ml/m^3)	(2) Number of observations	(3)* Preabatement mean volume (ml/m^3)	(4) Difference (1)–(3) (ml/m^3)
20	0.10	3	0.13	– 0.03
708	0.21	6	0.12	+ 0.09
710	0.14	7	0.20	– 0.06
608	0.19	6	0.22	– 0.03
605	0.20	6	0.27	– 0.07
16	0.19	7	0.19	0.00
401	0.38	6	0.35	+ 0.03
304	0.30	6	0.45	– 0.15
8	0.54	6	0.90	– 0.36
88	0.45	6	0.60	– 0.15
406	0.23	7	0.40	– 0.17

*From McNulty (1960).

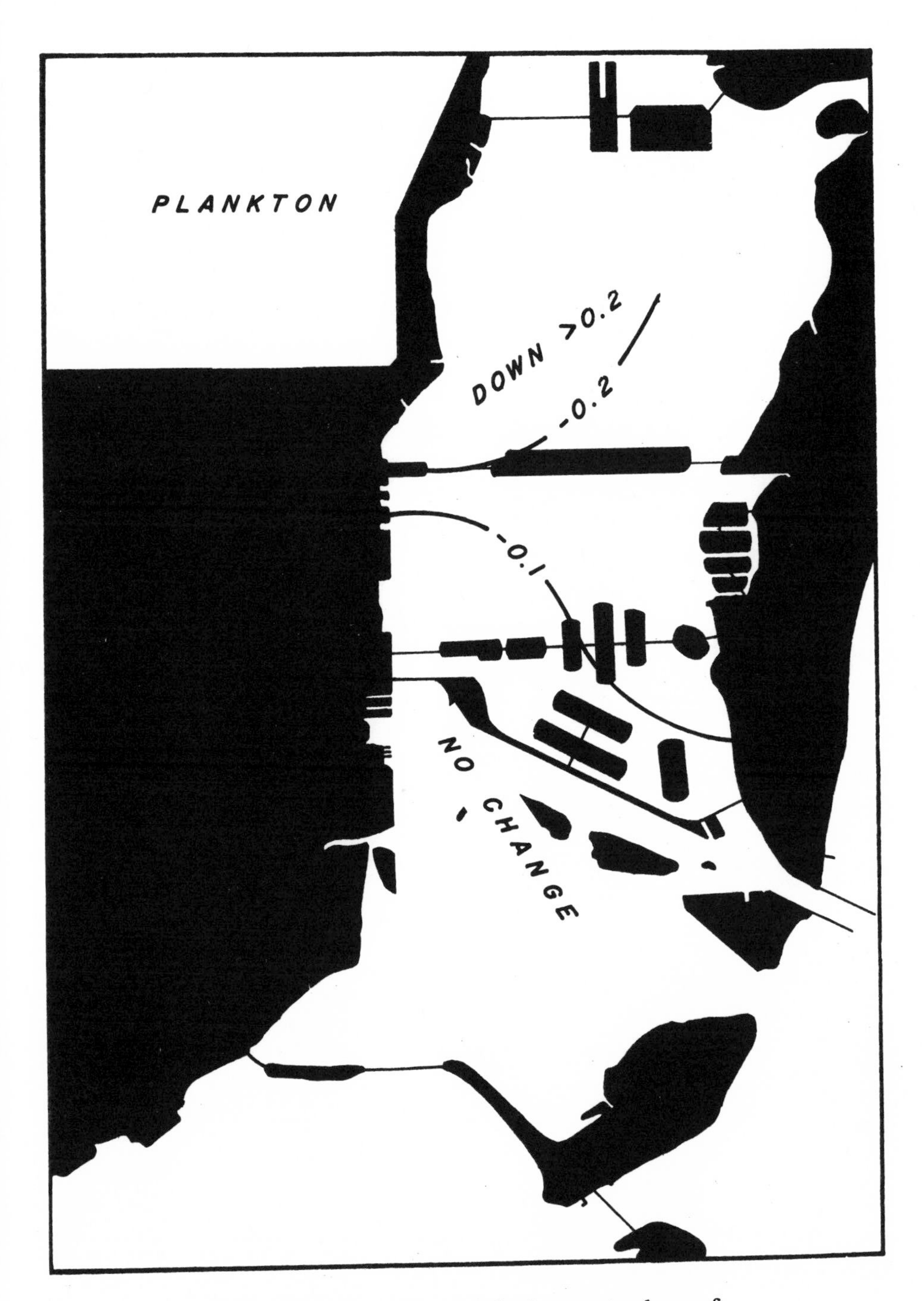

Fig. 16. Changes in the mean displacement volume of zooplankton in ml per m^3; minus sign indicates decrease in volumes after abatement of pollution

TABLE 8

Displacement Volumes of Fouling Organisms from Glass Panels Exposed One Month

Sta.	(1) Postabatement mean volume (ml/dm2)	(2) Number of observations	(3)* Preabatement mean volume (ml/dm2)	(4) Difference (1)–(3) (ml/dm2)
20	7.2	5	9.3	–2.1
708	6.5	5	4.9	+1.6
710	6.4	4	5.8	+0.6
608	3.1	5	0.7	+2.4
16	5.5	3	2.3	+3.2
401	8.8	5	12.1	–3.3
304	5.2	2	6.9	–1.7
8	7.4	5	10.7	–3.3
88	11.8	4	10.8	+1.0
406	5.6	4	2.8	+2.8

*From McNulty (1961).

TABLE 9

Abundance of Barnacles on Glass Panels Exposed One Month

Sta.	(1) Postabatement mean number (No /cm2)	(2) Number of observations	(3) Preabatement mean number (No /cm2)	(4) Difference (1)–(3) (No /cm2)
20	2.10	5	0.30	–1.80
708	1.61	5	0.85	+0.76
710	1.43	4	0.06	+1.37
608	0.88	5	0.20	+0.68
16	1.81	3	1.10	+0.71
401	2.84	5	0.23	+2.61
304	3.59	2	2.50	+1.09
8	2.19	5	0.44	+1.75
88	3.09	4	1.10	+1.99
406	1.24	4	1.44	–0.20

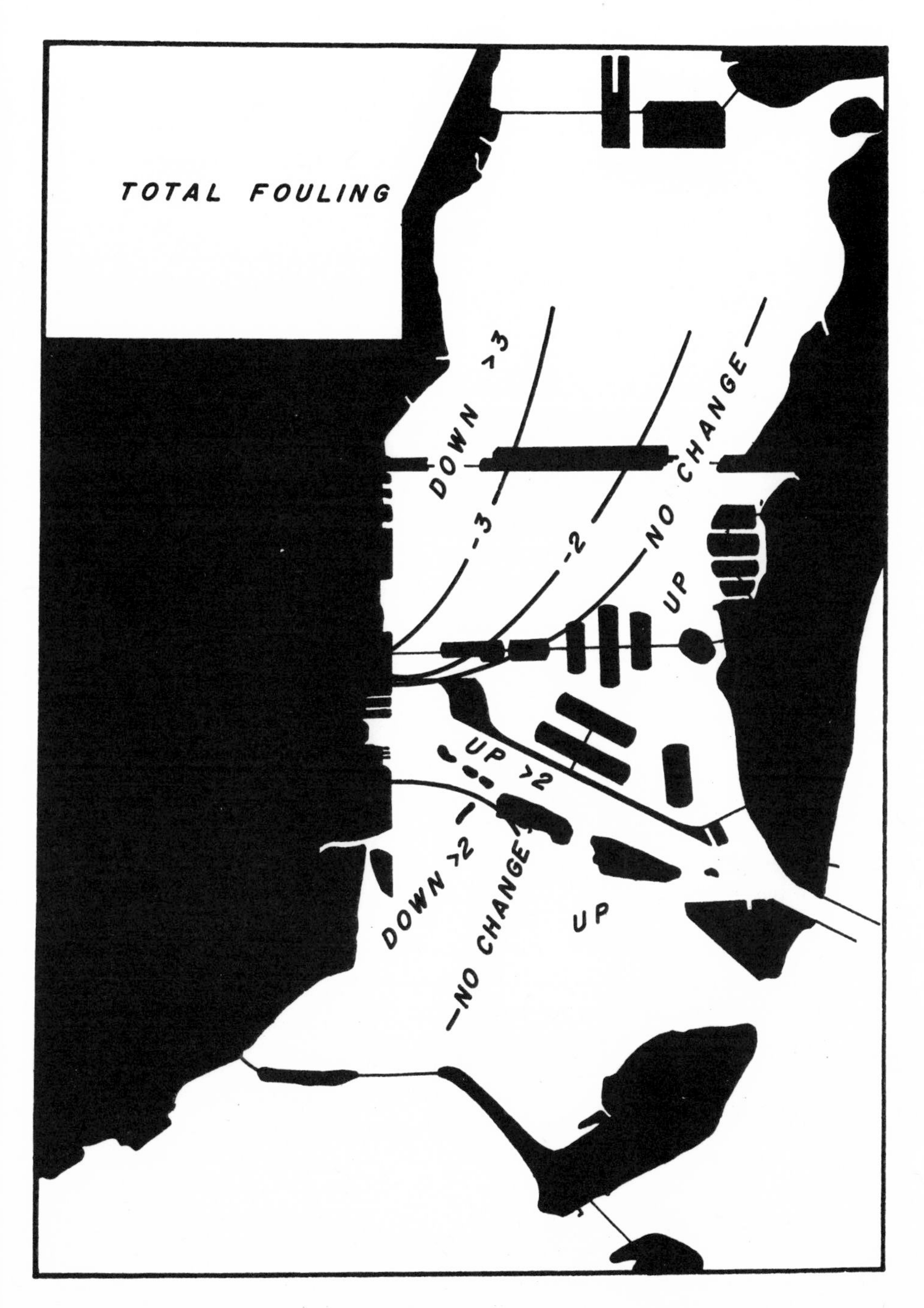

Fig. 17. Changes in mean displacement volume of fouling organisms settling on glass panels in one month, expressed as ml per dm^2; minus sign means volumes were smaller after abatement of pollution than before abatement

TABLE 10

Abundance of Amphipod Tubes on Glass Panels Exposed One Month

Sta.	(1) Postabatement mean number (No /cm2)	(2) Number of observations	(3)* Preabatement mean number (No /cm2)	(4) Difference (1)-(3) (No /cm2)
20	2.68	5	1.38	+1.30
708	1.70	5	0.02	+1.68
710	2.18	4	0.24	+1.94
608	0.44	5	0.04	+0.40
16	1.79	3	2.87	-1.08
401	3.42	5	4.03	-0.61
304	0.00	2	1.21	-1.21
8	1.91	5	2.11	-0.20
88	5.34	4	2.58	+2.76
406	1.43	4	0.31	+1.12

*From McNulty (1961).

Thus, there appears to be a clear relationship between the abundance of amphipods and pollution not shared by the barnacles. It is probably for this reason that when the two are considered together, as in total displacement volumes of fouling organisms, no clear relationship to pollution emerges.

It has been found in two other studies of estuarine pollution that *Balanus improvisus* Darwin is pollution-tolerant (Dean & Haskin, 1964; Filice, 1958). Since this species is the commonest barnacle in Biscayne Bay (Moore & Frue, 1959), and since together with *B. eburneus* Gould and *B. amphitrite niveus* Darwin it comprises one of the three species taken in highly polluted Biscayne Bay waters (McNulty, 1957), the results herein reported parallel the findings from San Francisco and Raritan bays previously cited. However, in Biscayne Bay it is *B. eburneus* that penetrates farther into fresh water than the other barnacles, whereas in the other two bays it appears that *B. improvisus* does so. Thus, the ecological niche occupied by one species in Biscayne Bay is occupied by another species in two other bays in different climates, and each species tolerates a similar combination of low salinity and high pollution. The situation is akin to that described by Thorson (1957), in which similar species replace one another in bottom communities in different parts of the world.

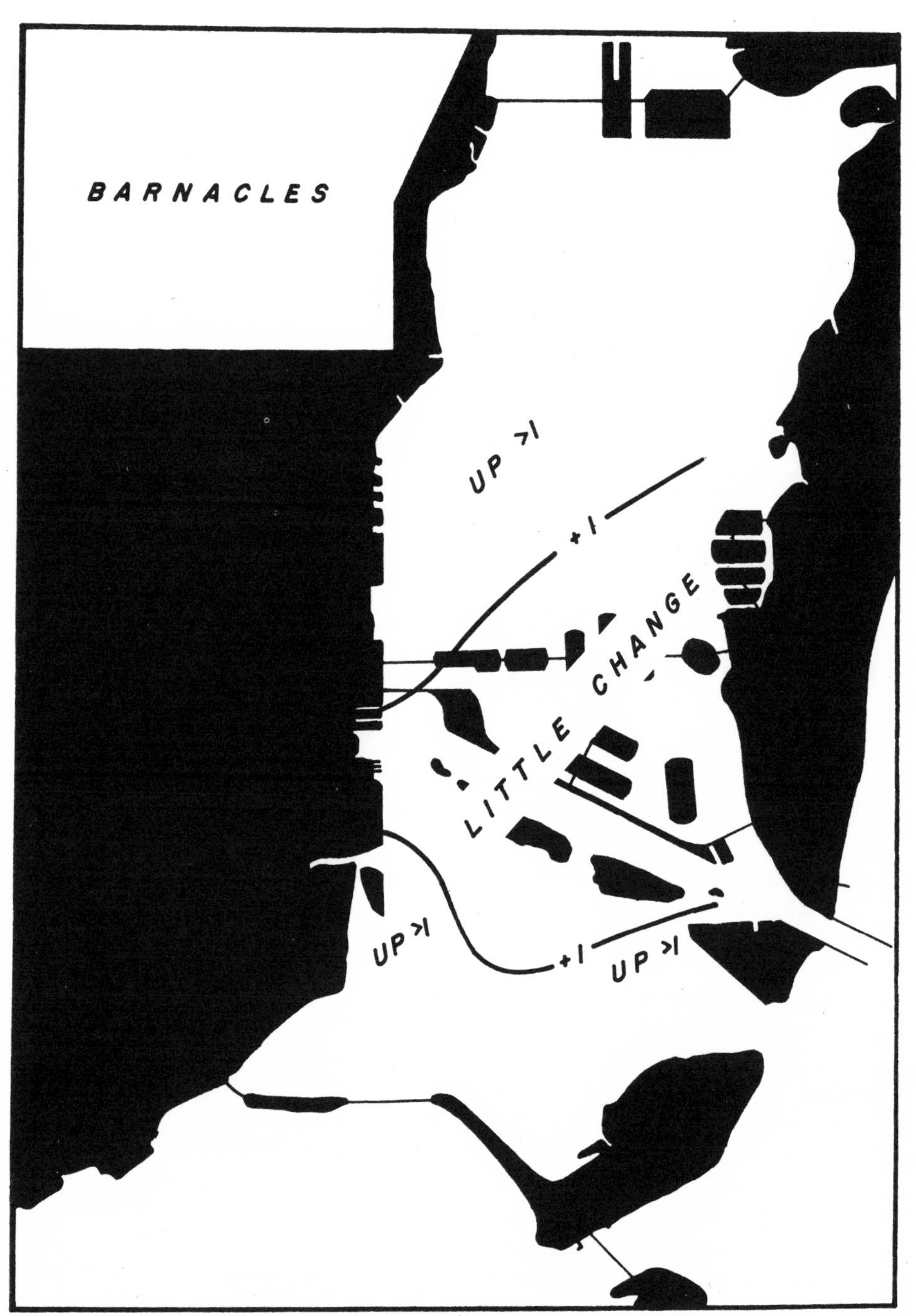

Fig. 18. Changes in the number of barnacles per cm^2 per month settling on glass panels exposed before and after abatement of pollution; plus sign means barnacles were more abundant after abatement of pollution than before abatement

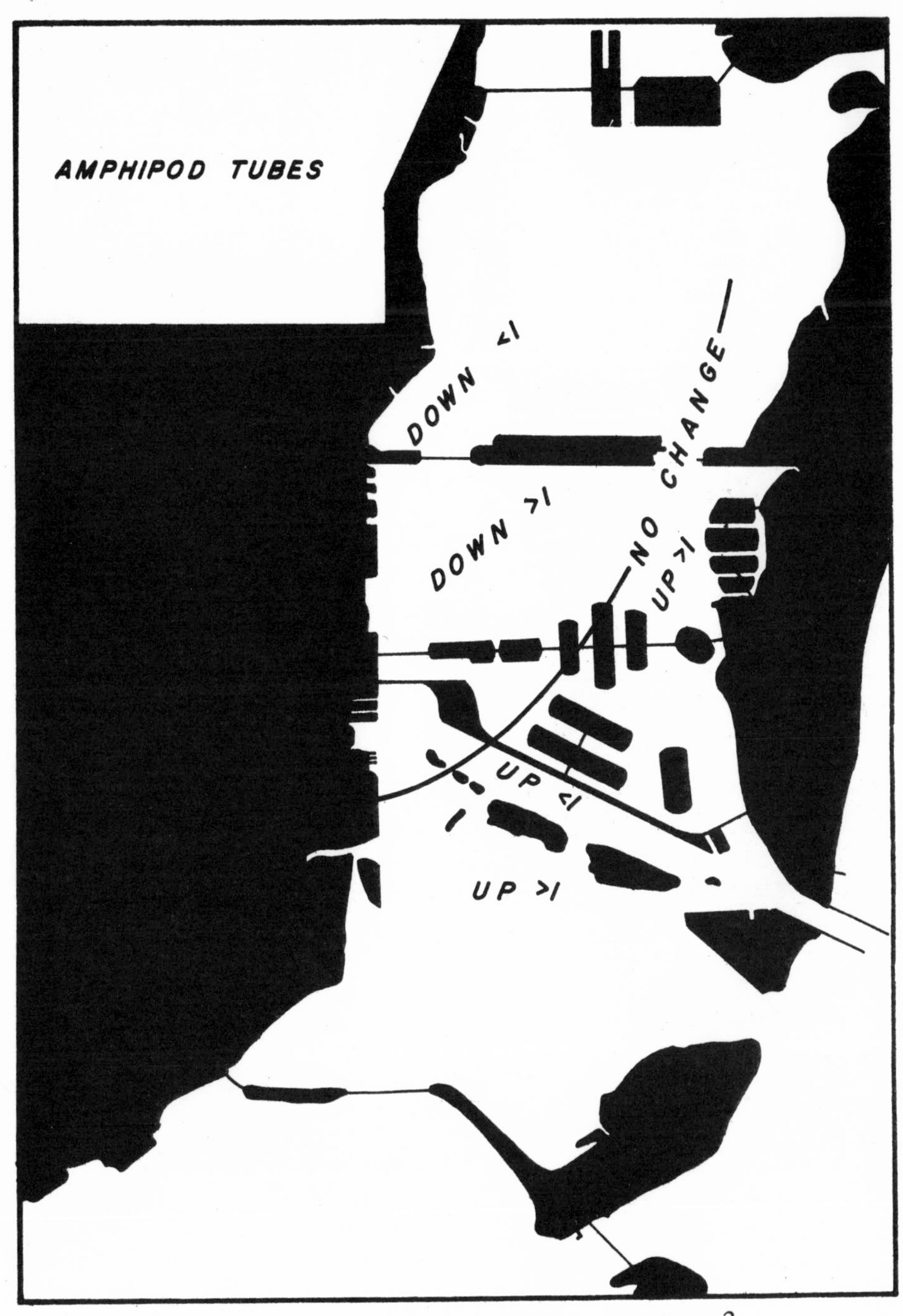

Fig. 19. Changes in the number of amphipod tubes per cm^2 per month on glass panels exposed before and after abatement of pollution

Discussion

Feeding types of the common benthic species can be determined, although in some cases the animals undoubtedly make use of a variety of methods. For this reason, a firm classification cannot be entirely true. Nevertheless, a list has been prepared based upon the following references: Blegvad (1915), Dales (1963), Hyman (1955), Hunt (1925), Jones (1963), MacGinitie & MacGinitie (1949), Nicol (1960), and Sanders (1956). The "characterizing species" and "immigrants" of the bottom communities are listed separately in an attempt to show possible pollution-related feeding tendencies, since the "immigrants" are representative of preabatement conditions.

Characterizing Species	Immigrants
Nonselective Deposit Feeders	Nonselective Deposit Feeders
Amphioplus	*Naineris*
Amphipholis	*Maldane*
Ophionephthys	
Suspension Feeders	Suspension Feeders
Codakia	*Corophium*
Chione	*Grubia*
Cyclinella	*Branchiomma*
Laevicardium	
Selective Deposit Feeders	Selective Deposit Feeders
Macoma	*Pista*
Nucula	*Polydora*
Tagelus	*Terebellides*
Tellina	
Cistenides	
Glycera	

Characterizing Species	Immigrants
Carnivores	Carnivores
Lumbrineris	*Bulla*
	Haminoea
	Onuphis
	Sthenelais
	Omnivorous Facultative Scavenger
	Nassarius

From the above, it appears that the carnivores and the omnivore were preferentially benefitted by pollution, at least in an environment with sufficient oxygen and with an abundance of prey, as in ecological zone "C" of the preceding section.

The autecologies of *Chione, Tagelus,* and *Nassarius* have been studied at this institution by H. B. Moore & N. N. López (1969), T. H. Fraser (1967), and R. H. Gore, respectively. The stuay by Gore is an unpublished paper for a course in marine ecology. The studies of *Chione* and *Tagelus* include observations on spawning periodicity and growth. Gore's study of *Nassarius* includes some growth data for a part of the year and many interesting observations on behavior. The latter observations have now been further expanded and published (Gore, 1966). Studies elsewhere include that of Klawe & Dickie (1957), which treats the embryology, feeding, growth, and population dynamics of *Glycera dibranchiata* in the Maritime Provinces.

Sanders (1956) has studied the growth of *Cistenides gouldii* in Long Island Sound, and Watson (1928) has worked out the life history of a related species, *Pectinaria (Lagis) koreni.* The only local benthic species on which there is an extensive literature is *Nassarius vibex.* Many papers on its embryology, behavior, feeding, growth, locomotion, parasitology, and thermal tolerance are available. The two aspects of its biology most pertinent to pollution are its tolerance of low oxygen and its omnivorous feeding ability. Although its tolerance of conditions low in content of dissolved oxygen is well known in general, its precise tolerance limits are apparently unknown. A large number of references to its feeding habits have been reviewed by Gore (1965).

Embryological studies of interest include those on *Glycera dibranchiata* by Stimpson (1962) and Klawe & Dickie (1957), and on *Diopatra cuprea* by Allen (1959). In addition, data on species related to local forms were provided by Costello *et al.* (1957), who included *Nucula proxima truncula, Sabellaria vulgaris,* and *Sthenelais leidyi. Polydora antennata* has been studied recently by Rullier (1963).

In summary, the life history data on our commonest benthic invertebrates are definitely scanty. A true understanding of their ecological relationships awaits more detailed investigation, species by species.

In sediments of similar nature, neither the scarcity of bottom fauna typical of areas close to sewer outfalls nor the tremendous abundance of life on the periphery of polluted areas has been encountered elsewhere in extensive sampling throughout Biscayne Bay and adjacent waters. The pollution acted as both depressant and stimulant on the benthic fauna, depending upon conditions.

The mechanism responsible for these effects can be described readily in its broad outlines. A rich suspension of organic material is intruded suddenly into the marine environment. The bacterial, fungal, and protozoan biota proliferates rapidly as it utilizes the huge store of food provided in great initial excess. In so doing, the dissolved oxygen is rapidly reduced to low concentrations, and the water becomes charged with the metabolic products of the saprophytes. A few tolerant species survive in small numbers, as seen near the outfalls in both hard and soft bottoms.

However, in an environment characterized by vigorous tidal currents sweeping across shallow areas, the resupply of dissolved oxygen is rapid, and septic conditions are quickly mitigated as the organic material is mixed and diluted with cleaner water. In the peripheral zone of pollution, conditions become favorable for a few of the more tolerant forms, such as certain amphipods, polychaetes, and lamellibranchs. They respond in this food-rich environment with vigorous growth in population. The proliferation goes unchecked by normal predators, such as certain crustaceans, echinoderms, and fishes, which are excluded by the still unnatural conditions. The result is the unusually dense concentration of infauna observed near the periphery of heavy pollution.

Eventually, and in the case of Biscayne Bay very rapidly, the mitigation of adverse conditions permits a return to normal diversity of species and abundance of individuals.

The general pattern is identical with that observed in other studies of the biological effects of pollution. It is unique in that every body of water has its own characteristic combination of hydrographic conditions and assemblage of species. Because of the relative nearness of the sea to the freshwater source in this estuary, and the fact that freshwater runoff is rapidly mixed with waters of high salinity, the mitigating effects of tidal action occur very soon after pollution is introduced. The resulting rapid recovery of the benthic fauna results in wide diversity of species and great abundance of individuals in firm sediment, above which there is rich food, ample oxygen, and vigorous water movement.

The author is firmly convinced that there was no benefit to man from the pollution within the study area, despite certain unusual concentrations of life described above. The polluted areas were conspicuously avoided by commercial and sport fishermen. The author made a few exploratory trawls in polluted areas during early planning stages of the preabatement study and found that the yield of fishes was meager, both qualitatively and quantitatively (McNulty, 1955). Partly because of the meager catches, but mostly for lack of time, fishery aspects were not pursued in this study. The few exploratory trawls did leave a lasting impression on the author, however, now confirmed by similar exploratory work in numerous bays of Florida over the past four years. The foregoing remarks are intended to dispel any false assumptions that the fertilizing effect of the pollution was beneficial to the fisheries. On the contrary, the available evidence indicates that the unusual concentrations of life described above went unutilized by sport and commercial fishes.

Summary

1. A domestic sewage disposal plant built by the City of Miami went into operation between September 18 and December 15, 1956, with the result that some 30 to 50 million gallons of waste per day were discharged into the disposal plant instead of into northern Biscayne Bay.
2. Ecological field studies by the author and others were conducted in the preabatement period between November, 1953, and September, 1956. Many of the field observations were repeated between November, 1960, and July, 1961. Results of the re-study are reported here, and comparisons of ecological conditions before and after abatement are made.
3. The study includes comparison of preabatement with postabatement distribution and abundance of benthic macroinvertebrates, concentration of dissolved inorganic phosphate-phosphorus, displacement volumes of zooplankton, numbers of barnacles, numbers of amphipod tubes, and volumes of all fouling organisms.
4. The effect of domestic sewage on the benthic macroinvertebrates is analyzed by attempting to show the changes in the clean-water bottom communities brought on by pollution. Thorson's method of naming bottom communities is used (Thorson, 1957).
5. Two bottom communities are described, one in soft bottom and the other in hard bottom. Their major characterizing species are, respectively, the ophiuran, *Ophionephthys limicola,* and the venerid lamellibranch, *Chione cancellata.*
6. Pollution affected the soft-bottom community less than it affected the hard-bottom community, possibly because animals of the soft sediment are adapted to higher organic content of the environment than are animals in hard bottom.

7. Near outfalls, in relatively deep water and soft bottom under polluted conditions, there were few species and few individuals per species. Under clean water conditions, the species of polychaetes were different, but the fauna remained impoverished both qualitatively and quantitatively.
8. A few hundred meters seaward from outfalls, in soft bottom under polluted conditions, about half of the normally common species were missing, while several immigrant species normally not in the community were present. The total quantity of animals before and after abatement remained about the same.
9. Near an outfall, in shallow water and hard bottom under polluted conditions, the fauna was depressed. After pollution abatement, it increased markedly both qualitatively and quantitatively.
10. A few hundred meters seaward from outfalls, in shallow water and hard bottom under polluted conditions, the fauna was unusually diverse and abundant. Many more species were found and were at much higher population levels than under any other conditions.
11. The diversity and abundance of attached vegetation declined between 1956 and 1960, contrary to expectations. It is possible that turbidity and sedimentation from hydraulic dredging for construction of the Julia Tuttle Causeway may have been a major factor in the decline. Another event of possible significance was Hurricane Donna, which, on October 10, 1960, brought sustained easterly winds of 89-101 kilometers per hour with gusts to 129-132 kilometers per hour.
12. The concentration of dissolved inorganic phosphate-phosphorus dropped markedly following abatement of pollution. Mean concentrations in Miami River water dropped to about half the preabatement levels, while decreases to about one-quarter the preabatement level occurred in the northwestern part of the study area.
13. Volumes of zooplankton decreased to about half the mean volumes recorded in preabatement work in the northwestern part of the study area. In other parts of the study area, however, volumes remained about the same, indicating that pollution-related factors stimulated production of zooplankton only in the poorly flushed, nutrient-rich, northwestern study area.
14. The abundance of amphipod tubes declined markedly following abatement of pollution, as shown by the monthly settlement of fouling organisms on glass panels exposed throughout the study area. Neither the total displacement volume of all fouling organisms nor the numbers of barnacles settling per month show a clear relationship to pollution.
15. The literature on pollution in the marine environment is reviewed.
16. A summary of the geology of Biscayne Bay is presented.
17. The history of dredging and filling in the area of interest is reviewed briefly.

Abstract

Various elements of the biota of northern Biscayne Bay, Florida, were studied before and after abatement of pollution. The pollution consisted of 136 to 227 million liters per day of untreated domestic sewage. Four years after removal of the pollution certain changes had taken place. At distances of 100 to 740 meters seaward from outfalls, in water depths of one to three meters in hard bottom, populations of benthic macroinvertebrates had declined from abnormally large numbers of species and individuals to normal numbers of each, while soft-bottom populations had changed qualitatively but not quantitatively. Adjacent to outfalls, populations had increased in numbers of species and numbers of individuals in hard sandy bottoms only. Volumes of zooplankton had decreased to about one-half the pre-abatement values in poorly flushed waters; elsewhere, they remained about the same. Dissolved inorganic phosphate-phosphorus decreased similarly. Abundance of amphipod tubes had declined markedly, a change not shared by the quantities of other fouling organisms (including barnacles), which remained about the same. There was no evidence of improved commercial and sport fishing following abatement; this is interpreted to mean that long-lasting detrimental effects have resulted from pollution and dredging.

Bibliography

ALEXANDER, W. B., B. A. SOUTHGATE, and R. BASSINDALE
1936. Summary of Tees estuary investigations. Survey of the River Tees. Part II. The estuary, chemical and biological. J. mar. biol. Ass. U.K., *20* (3): 717-724.

ALLEN, M. J.
1959. Embryological development of the polychaetous annelid, *Diopatra cuprea* (Bosc). Biol. Bull. mar. biol. Lab., Woods Hole, *116* (3): 339-361.

BAKER, F. C.
1926. The changes in the bottom fauna of the Illinois River due to pollutional causes. Ecology, 7 (2): 229-230.

BARLOW, J. P., C. J. LORENZEN, and R. T. MYREN
1963. Eutrophication of a tidal estuary. Limnol. Oceanogr., *8* (2): 251-262.

BARNARD, J. L.
1958. Amphipod crustaceans as fouling organisms in Los Angeles-Long Beach harbors, with reference to the influence of seawater turbidity. Calif. Fish Game, *44:* 161-170, 2 figs.

BARNARD, J. L. and D. J. REISH
1959. Ecology of Amphipoda and Polychaeta of Newport Bay, California. Occ. Pap. Allan Hancock Fdn., No. 21, 106 pp., 14 pls.

BARTSCH, A. F.
1948. Biological aspects of stream pollution. Sewage Wks. J., *20* (2): 292-302.

BARTSCH, A. F. and W. S. CHURCHILL
1949. Biotic responses to stream pollution during artificial stream reaeration. Pp. 33-48 *in* F. R. Moulton and F. Hitzel (Eds.), Limnological aspects of water supply and waste disposal. Amer. Assoc. Adv. Sci., Washington, D.C., 87 pp.

BASSINDALE, R.
1938. The intertidal fauna of the Mersey estuary. J. mar. biol. Ass. U.K., *23* (1): 83-98.

BASU, A. K.
1965. Observations on the probable effects of pollution on the primary productivity of the Hooghly and Mutlah estuaries. Hydrobiologia, *25* (1-2): 302-316.

BAUGHMAN, J. L.
1948. An annotated bibliography of oysters with pertinent materials on mussels and other shellfish and an appendix on pollution. The Texas A & M Research Foundation, College Station, Texas, 794 pp.

BEHRE, KARL
1963. Die Algenbesiedlung einiger Häfen in Bremerhaven und ihre Beziehungen zur Verschmutzung dieser Gewässer. Veröff. Inst. Meeresforsch. Bremerh., *8* (2): 192-249. (Illus.)

BELLAN, G.
1964. Contribution à l'étude systématique, bionomique et écologique des annélides polychètes de la Méditerranée. Rec. Trav. Sta. mar. d'Endoume, Bull., *33* (49), 371 pp.

BLEGVAD, H.
1915. Food and conditions of nourishment among the communities of invertebrate animals found on or in the sea bottom in Danish waters. Rep. Danish biol. Sta., No. 22: 41-78.
1932. Investigations of the bottom fauna at outfalls of drains in the Sound. Rep. Danish biol. Sta., No. 37: 1-20, 4 figs.

BRINKHURST, R.O.
1965. Observations on the recovery of a British river from gross organic pollution. Hydrobiologia, *25* (1-2): 9-51.

BUTLER, P. A. and P. F. SPRINGER
1963. Pesticides–a new factor in coastal environments. Trans. N. Am. Wildl. Conf., *28:* 378-390.

CALIFORNIA STATE WATER POLLUTION CONTROL BOARD

1956. An investigation of the efficacy of submarine outfall disposal of sewage and sludge. Publs Calif. St. Wat. Pollut. Control Bd., No. 14, 154 pp.

1960. Summary of marine waste disposal research program in California. Publs Calif. St. Wat. Pollut. Control Bd., No. 22, 77 pp.

1964a. An oceanographic study between points of Trinidad Head and the Eel River. Publs Calif. St. Wat. Pollut. Control Bd., No. 25, xii + 135 pp.

1964b. An investigation of the effects of discharged wastes on kelp. Publs Calif. St. Wat. Pollut. Control Bd., No. 26, xiv + 124 pp.

1965a. An oceanographic and biological survey of the southern California mainland shelf. Publs Calif. St. Wat. Pollut. Control Bd., No. 27, xiv + 232 pp.

1965b. An investigation on the fate of organic and inorganic water discharged into the marine environment and their effects on biological productivity. Publs Calif. St. Wat. Pollut. Control Bd., No. 29, xi + 117 pp.

CALIFORNIA STATE WATER QUALITY CONTROL BOARD

1963. Water quality criteria. J. E. McKee and H. W. Wolf (Eds.). Publs Calif. St. Wat. Qual. Control Bd., No. 3-A, xiv + 548 pp. (Bibliography of 3,827 entries.)

CARPENTER, K. E.

1924. A study of the fauna of rivers polluted by lead mining in the Aberystwyth district of Cardiganshire. Ann. appl. Biol., *11:* 1-23.

CARRUTHERS, J. N.

1954. Sea fish and residual sewage effluents. Intelligence Digest Supplement, page unknown. Coll. Repr. natn. Inst. Oceanogr.

CLARKE, GEORGE L. and D. F. BUMPUS

1940. The plankton sampler—an instrument for quantitative plankton investigations. Spec. Publs Limnol. Soc. Am., No. 5, 8 pp., 5 figs.

COSTELLO, D.P., M.E. DAVIDSON, A. EGGERS, M.H. FOX, and C. HENLEY

1957. Methods for obtaining and handling marine eggs and embryos. Lancaster Press, Lancaster, Pa., xv + 247 pp.

DADE COUNTY, FLORIDA. PLANNING DEPARTMENT

1960. Economic base study. 210 pp. (Unpublished.)

DALES, R. P.

1963. Annelids. Hutchinson & Co., London, 200 pp.

DAVIS, CHARLES C.
1950. Observations of plankton taken in marine waters of Florida in 1947 and 1948. Quart. J. Fla. Acad. Sci., *12* (2): 67-103.

DAVIS, JOHN H., JR.
1943. Natural features of southern Florida. Geol. Bull. Fla., No. 25, 311 pp.

DEAN, DAVID, and HAROLD H. HASKIN
1964. Benthic repopulation of the Raritan River estuary following pollution abatement. Limnol. Oceanogr., *9* (4): 551-563.

FILICE, FRANCIS P.
1954a. An ecological survey of the Castro Creek area in San Pablo Bay. Wasmann J. Biol., *12* (1): 1-24.
1954b. A study of some factors affecting the bottom fauna of a portion of the San Francisco Bay estuary. Wasmann J. Biol., *12* (3): 257-292.
1958. Invertebrates from the estuarine portion of San Francisco Bay and some factors influencing their distributions. Wasmann J. Biol., *16* (2): 159-211.
1959. The effect of wastes on the distribution of bottom invertebrates in the San Francisco Bay estuary. Wasmann J. Biol., *17* (1): 1-17.

FØYN, E.
1960. Chemical and biological aspects of sewage disposal in inner Oslofjord. Pp. 279-294 *in* E. A. Pearson (Ed.), Waste disposal in the marine environment. Pergamon Press, New York, 569 pp.

FRASER, J. H.
1932. Observations on the fauna and constituents of an estaurine mud in a polluted area. J. mar. biol. Ass. U.K., *18* (1): 69-85.

FRASER, THOMAS H.
1967. Contributions to the biology of *Tagelus divisus* (Tellinacea:Pelecypoda) in Biscayne Bay, Florida. Bull. Mar. Sci., 17 (1): 111-132.

GAUFIN, ARDEN R.
1957. The use and value of aquatic insects as indicators of organic enrichment. Pp. 136-143 *in* C. M. Tarzwell (Ed.), Biological problems in water pollution. U.S. Public Health Service, Robert A. Taft Sanitary Engineering Center, Cincinnati, Ohio, 272 pp.

GAUFIN, A. R. and C. M. TARZWELL
1952. Aquatic invertebrates as indicators of stream pollution. Publ. Hlth. Rep., Wash., *67:* 57-64.
1956. Aquatic macro-invertebrate communities as indicators of organic pollution in Lytle Creek. Sewage ind. Wastes, *28:* 906-924.

GILET, R.

1960. Water pollution in Marseilles and its relation with flora and fauna. Pp. 39-56 *in* E. A. Pearson (Ed.), Proceedings of the first international conference on waste disposal in the marine environment. Pergamon Press, New York, 569 pp.

GORE, ROBERT H.

1965. Aspects of the biology of *Nassarius vibex* (Say) with a literature survey of the genus *Nassarius.* Institute of Marine Sciences, University of Miami, unpublished MS.

1966. Observations on the escape response in *Nassarius vibex* (Say), (Mollusca: Gastropoda). Bull. Mar. Sci., *16* (3): 423-434.

GREENFIELD, L. J.

1952. The distribution of marine borers in the Miami area in relation to ecological conditions. Bull. Mar. Sci. Gulf. & Carib., *2* (2): 448-464.

GREENFIELD, L. J. and F. A. KALBER

1954. Inorganic phosphate measurement in seawater. Bull. Mar. Sci. Gulf & Carib., *4* (4): 323-335.

GROSS, F.

1949a. Further observations on fish growth in a fertilized sea loch (Loch Craiglin). J. mar. biol. Ass. U.K., *28:* 1-8.

1949b. Experiments in marine fish cultivation. XIII Congrès International de Zoologie: 392-393.

GROSS, F., J. E. G. RAYMONT, S. M. RAYMONT, and A. P. ORR

1944. A fish-farming experiment in a sea loch. Nature, *153:* 483. (Not seen.)

HARTMAN, OLGA

1955. Quantitative survey of the benthos of San Pedro Basin, southern California. Part I. Preliminary results. Allan Hancock Pacif. Exped., *19:* 1-185.

1960. The benthonic fauna of southern California in shallow depths and possible effects of wastes on the marine biota. Pp. 57-81 *in* E. A. Pearson (Ed.), Proceedings of the first international conference on waste disposal in the marine environment. Pergamon Press, New York, 569 pp.

HAWKES, H. A.

1962. Biological aspects of river pollution. Pp. 311-432 (Chap. 8) *in* Louis Klein, River pollution II. Causes and effects. Butterworths, London, xiv + 456 pp.

HELA, I., C. A. CARPENTER, JR., and J. K. McNULTY
1957. Hydrography of a positive, shallow, tidal bar-built estuary (report on the hydrography of the polluted area of Biscayne Bay). Bull. Mar. Sci. Gulf & Carib., *7* (1): 47-99.

HOHN, MATTHEW H.
1959. The use of diatom populations as a measure of water quality in selected areas of Galveston and Chocolate Bay, Texas. Publs Inst. mar. Sci. Univ. Tex., *6:* 206-212.

HOLLINGSWORTH, TRACY
1936. History of Dade County, Florida. Publisher not shown. 151 pp.

HOOD, DONALD W., W. DUKE, and B. STEVENSON
1960. Measurement of toxicity of organic wastes to marine organisms. J. Wat.Pollut. Control Fed., *32* (9): 982-993.

HUNT, O. D.
1925. The food of the bottom fauna of the Plymouth fishing grounds. J. mar. biol. Ass. U.K., *13* (3): 560-599.

HYMAN, L. H.
1955. The invertebrates: Echinodermata. McGraw-Hill, New York, vii + 763 pp.

HYNES, H. B. N.
1960. The biology of polluted waters. Liverpool University Press, Liverpool, xiv+202 pp.

HYPERION ENGINEERS
1957. Ocean outfall design. Holmes & Narver, Inc., Los Angeles, California, 851 pp.

INGRAM, W. M. and T. A. WASTLER, III
1961. Estuaries and marine pollution. U.S. Public Health Service, Robert A. Taft Sanitary Engineering Center, Cincinnati, Ohio. Tech. Report W-61-4, 30 pp.

INTERNATIONAL ATOMIC ENERGY AGENCY
1960a. Disposal of radioactive wastes. Vol. I. K. Ring, Vienna, 603 pp.
1960b. Disposal of radioactive wastes. Vol. II. K. Ring, Vienna, 575 pp.

JEFFRIES, H. P.
1962. Environmental characteristics of Raritan Bay, a polluted estuary. Limnol. Oceanogr., *7* (1): 21-31.

JONES, M. L.
1963. Complexities in the substrate. Nat. Hist., N.Y., *72* (5): 10-17.

JOSEPH, E. B. and F. E. NICHY
1955. Literature survey of the Biscayne Bay area. Part II. Algae, marine fouling and boring organisms. The Oceanographic Institute, Florida State University, Tallahassee, Fla., 33 pp. (Mimeographed.)

KETCHUM, B. H. and W. L. FORD
1952. Rate of dispersion in the wake of a barge at sea. Trans. Am. geophys. Un., *33* (5): 680-684.

KING, D. L. and R. C. BALL
1964. A quantitative biological measure of stream pollution. J. Wat. Pollut. Control Fed., *36* (5): 650-653.

KLAWE, W. L. and L. M. DICKIE
1957. Biology of the bloodworm, *Glycera dibranchiata* Ehlers, and its relation to the bloodworm fishery of the Maritime Provinces. Bull. Fish. res. bd. Can., No. 115, iii + 37 pp.

KOLKWITZ, R. and M. MARSSON
1908. Oekologie die pflanzlichen Saprobien. Ber. dt. bot. Ges., *26a:* 505-519.
1909. Oekologie der tierischen Saprobien. Int. Rev. Hydrobiol., *2:* 126-152.

LACKEY, J. B.
1960. The status of plankton determination in marine pollution analysis. Pp. 404-412 *in* E. A. Pearson (Ed.), Proceedings of the first international conference on waste disposal in the marine environment. Pergamon Press, New York, 569 pp.

LAURIE, R. D. and J. R. E. JONES
1938. The faunistic recovery of a lead-polluted river in north Cardiganshire, Wales. J. Animal Ecol., *7:* 272-289.

LIEBMAN, E.
1940. River discharges and their effect on the cycles and productivity of the sea. Proc. Sixth Pacif. Sci. Congr., *3:* 517-523.

LYNN, W. A. and W. T. YANG
1960. The ecological effects of sewage in Biscayne Bay. Oxygen demand and organic carbon determinations. Bull. Mar. Sci. Gulf & Carib., *10* (4): 491-509.

MacGINITIE G.E.
1939. Some effects of fresh water on the fauna of a marine harbor. Amer. Midl. Nat., *21:* 681-686.

MacGINITIE G.E. and N. MacGINITIE
1949. Natural history of marine animals. McGraw-Hill, New York, xii + 473 pp., 282 figs.

MACKENTHUM, K. M.
1965. Nitrogen and phosphorus in water. U.S. Public Health Service, Government Printing Office, Washington, D.C., 111 pp.

MACKENTHUM, K. M. and W. M. INGRAM
1964. Limnological aspects of recreational lakes. U.S. Public Health Service, Government Printing Office, Washington, D.C., 176 pp.

MARTENS, J. H. C.
1935. Beach sands between Charleston, South Carolina, and Miami, Florida. Bull. geol. Soc. Am., *46:* 1563-1596.

MAUCHLINE, J. and W. L. TEMPLETON
1964. Artificial and natural radioisotopes in the marine environment. Oceanogr. Mar. Biol., *2:* 229-279.

McNULTY, J. K.
1955. Macroorganism studies. Pp. IV-1 through IV-6 *in* Moore, H. B., I. Hela, E. S. Reynolds, J. K. McNulty, S. Miller, and C. A. Carpenter, Jr., Report on preliminary studies of pollution in Biscayne Bay. Progress report to National Institutes of Health. The Marine Laboratory, University of Miami, Coral Gables, Fla., Mimeographed Report Series, No. 55-3, 78 pp. (Unpublished MS.)
1957. Pollution studies in Biscayne Bay during 1965. Progress report to National Institutes of Health. The Marine Laboratory, University of Miami, Coral Gables, Fla., Mimeographed Report Series, No. 57-8, 15 pp., 11 figs. (Unpublished MS.)
1961. Ecological effects of sewage pollution in Biscayne Bay, Florida: sediments and the distribution of benthic and fouling macroorganisms. Bull. Mar. Sci. Gulf & Carib., *11* (3): 394-447.

McNULTY, J. K., E. S. REYNOLDS, and S. M. MILLER
1960. Ecological effects of sewage pollution in Biscayne Bay, Florida: distribution of coliform bacteria, chemical nutrients, and volume of zooplankton. Pp. 189-202 *in* C. M. Tarzwell (Compiler), Biological problems in water pollution. U.S. Public Health Service, Robert A. Taft Sanitary Engineering Center, Cincinnati, Ohio. Tech. Rept. W 60-3, 285 pp.

McNULTY, J. K., R. C. WORK, and H. B. MOORE

1962a. Level sea bottom communities in Biscayne Bay and neighboring areas. Bull. Mar. Sci. Gulf & Carib., *12* (2): 204-233.

1962b. Some relationships between the infauna of the level bottom and the sediment in south Florida. Bull. Mar. Sci. Gulf & Carib., *12* (3): 322-332.

MEYERS, S. P.

1953. Marine fungi in Biscayne Bay, Florida. Bull. Mar. Sci. Gulf & Carib., *2* (4): 590-601.

1954. Marine fungi in Biscayne Bay, Florida. II. Further studies of occurrence and distribution. Bull. Mar. Sci. Gulf & Carib., *3* (4): 307-327.

MILLIKEN, D. L.

1949. Report on investigation of water resources of Biscayne Bay, Florida, May-August, 1949. U.S. Geological Survey in cooperation with City of Miami, Florida, 71 pp. (Typewritten.)

MINKIN, J. L.

1949. Biscayne Bay pollution survey, May-October, 1949. Florida State Board of Health, Bureau of Sanitary Engineering, Jacksonville, Florida, 78 pp. (Mimeographed.)

MOHR, J. L.

1953. The relationship of the areas of marine borer attack to pollution patterns in Los Angeles-Long Beach harbors. Pp. I-1 through I-5 *in* Marine borer conference, Miami Beach, Fla., 1952. The Marine Laboratory, Univ. of Miami, 144 pp. (Mimeographed.)

1960. Biological indicators of organic enrichment in marine habitats. Pp. 237-239 *in* C. M. Tarzwell (Compiler), Biological problems in water pollution. U.S. Public Health Service, Robert A. Taft Sanitary Engineering Center, Cincinnati, Ohio. Tech. Rept. W60-3, 285 pp.

MOORE, D. R.

1963. Distribution of the sea grass, *Thalassia,* in the United States. Bull. Mar. Sci. Gulf & Carib., *13* (2): 329-342.

MOORE, H. B. and A. C. FRUE

1959. The settlement and growth of *Balanus improvisus, B. eburneus* and *B. amphitrite* in the Miami area. Bull. Mar. Sci. Gulf & Carib., *9* (4): 421-440.

MOORE, H. B. and N. N. LÓPEZ

1969. The ecology of *Chione cancellata.* Bull. Mar. Sci., *19* (1): 131-148.

MORRILL, J. B., JR. and F. C. W. OLSON
1955. Literature survey of the Biscayne Bay area. The Oceanographic Institute, Florida State University, Tallahassee, Fla., 134 pp. (Mimeographed.)

MUNROE. R. M. and V. GILPIN
1930. The commodore's story. Ives Washburn, New York, xv + 384 pp.

MURPHY, J. and J. P. RILEY
1958. A single-solution method for the determination of soluble phosphate in seawater. J. mar. biol. Ass. U.K., *37* (1): 9-14.

NATIONAL RESEARCH COUNCIL
1957. The effects of atomic radiation on oceanography and fisheries. National Academy of Sciences, Washington, D.C., Publ. No. 551, ix + 137 pp.
1959. Radioactive waste disposal into Atlantic and Gulf coastal waters. National Academy of Sciences, Washington, viii + 37 pp.

NELSON, T. C.
1960. Some aspects of pollution, parasitism and inlet restriction in three New Jersey estuaries. Pp. 203-211 *in* C. M. Tarzwell (Compiler), Biological problems in water pollution. U.S. Public Health Service, Robert A. Taft Sanitary Engineering Center, Cincinnati, Ohio. Tech. Rept. W60-3, 285 pp.

NEWELL, G. E.
1959. Pollution and the abundance of animals in estuaries. Pp. 61-69 *in* W. G. Yapp (Ed.), The effects of pollution on living material. The Institute of Biology, London, 154 pp.

NICOL, J. A. C.
1960. The biology of marine animals. Pitman & Sons, London, xi + 707 pp.

ODUM, E. P.
1963. Ecology. Holt, Rinehart and Winston, New York, 152 pp.

ODUM, H. T.
1960. Analysis of diurnal oxygen curves for the assay of reaeration rates and metabolism in polluted marine bays. Pp. 547-555 *in* E. A. Pearson (Ed.), Waste disposal in the marine environment. Pergamon Press, New York, 569 pp.

ODUM, H. T., R. P. C. DUREST, J. BEYERS, and C. ALLBAUGH
1963. Diurnal metabolism, total phosphorus, Ohle anomaly, and zooplankton diversity of abnormal marine ecosystems of Texas. Publs Inst. mar. Sci. Univ. Tex., *9*:404-453.

O'GOWER, A. K. and J. W. WACASEY

1967. Animal communities associated with *Thalassia, Diplanthera,* and sand beds in Biscayne Bay. I. Analysis of communities in relation to water movements. Bull. Mar. Sci., *17* (1): 175-201.

PARKER, G. G., N. D. HOY, and M. C. SCHROEDER

1955. Relationship of geology to study of groundwater resources. Pp. 57-125 *in* G. G. Parker, G. E. Ferguson, S. K. Love, and others, Water resources of southeastern Florida. U.S. Geological Survey water-supply paper No. 1255, Government Printing Office, Washington, D.C., 965 pp.

PATRICK, R.

1949. A proposed biological measure of stream conditions, based on a survey of the Conestoga Basin, Lancaster County, Pennsylvania. Proc. Acad. nat. Sci. Philad., *101:* 227-341, 77 figs.

PATRICK, RUTH, M. H. HOHN, and J. H. WALLACE

1954. A new method for determining the pattern of the diatom flora. Notulae Naturae, No. 259, 12 pp.

PETERSEN, C. G. J.

1918. The sea bottom and its production of fish-food. A survey of the work done in connection with the valuation of the Danish waters from 1883-1917. Rep. Dan. biol. Stn., No. *25,* 62 pp., 10 pls., 1 chart.

PRITCHARD, D. W.

1952. A review of our present knowledge of the dynamics and flushing of estuaries. Chesapeake Bay Institute, The Johns Hopkins University. Tech. Rept. 4, Reference 52-7, 45 pp. (Mimeographed.)

REISH, D. J.

1955. The relation of polychaetous annelids to harbor pollution. Publ. Hlth Rep., Wash., *70:*1168-1174.

1956. An ecological study of lower San Gabriel River, California, with special reference to pollution. Calif. Fish Game, *42* (1): 51-61.

1957a. Effect of pollution on marine life. Ind. Wastes, *2:* 114-118.

1957b. The relationship of the polychaetous annelid *Capitella capitata* (Fabricius) to waste discharges of biological origin. Pp. 195-200 *in* C. M. Tarzwell (Ed.), Biological problems in water pollution. U.S. Public Health Service, Robert A. Taft Sanitary Engineering Center, Cincinnati, Ohio, 272 pp.

1959. An ecological study of pollution in Los Angeles-Long Beach harbors, California. Occ. Pap. Allan Hancock Fdn., No. 22, 119 pp., 18 pls.

1960. The use of marine invertebrates as indicators of water quality. Pp. 92-103 *in* E. A. Pearson (Ed.), Waste disposal in the marine environment. Pergamon Press, New York, 569 pp.

1961. The use of the sediment bottle collector for monitoring polluted marine waters. Calif. Fish Game, *47* (3): 261-272.

REISH, D. J. and J. L. BARNARD
1960. Field toxicity tests in marine waters utilizing the polychaetous annelid *Capitella capitata* (Fabricius). Pacif. Nat., *1* (21): 1-8.

REISH, D. J. and H. A. WINTER
1954. The ecology of Alamitos Bay, California, with special reference to pollution. Calif. Fish Game, *40* (2): 105-121.

RICHARDSON, ROBERT E.
1921. Changes in the bottom and shore fauna of the middle Illinois River and its connecting lakes since 1913-1915 as a result of the increase, southward, of sewage pollution. Bull. Ill. St. nat. Hist. Surv., *14:* 33-75.
1928. The bottom fauna of the middle Illinois River, Illinois, 1913-1925. Bull. Ill. St. nat. Hist. Surv., *17:* 387-475.

RILEY, G. A.
1937. The significance of the Mississippi River drainage for biological conditions in the northern Gulf of Mexico. J. mar. Res., *1* (1): 60-74.

ROBINSON. R. J. and T. G. THOMPSON
1948. The determination of phosphates in sea water. J. mar. Res., *7* (1): 33-41.

RULLIER, F.
1963. Développement de *Polydora (Carazzia) antennata* Clap. var. *pulchra* Carazzi. Cahiers Biol. Mar., *4:* 233-250.

RYTHER, J. H.
1954. The ecology of phytoplankton blooms in Moriches Bay and Great South Bay, Long Island, New York. Biol. Bull. mar. biol. Lab., Woods Hole, *106* (2): 198-209.

RYTHER, J. H., C. S. YENTSCH, E. M. HULBURT, and R. F. VACCARO
1958. The dynamics of a diatom bloom. Biol. Bull. mar. biol. Lab., Woods Hole, *115* (2): 257-268.

SADDINGTON, K. and W. L. TEMPLETON
1958. Disposal of radioactive waste. G. Newnes, London, x + 102 pp.

SANDERS, H. L.
1956. Oceanography of Long Island Sound, 1952-1954. X. The biology of marine bottom communities. Bull. Bing. Oceanog. coll., *15:* 345-414.

SCHLIENZ, W.
1923. Verbreitung und Verbreitungsbedingungen der höheren Krebse im Mündungsgebiet der Elbe. Arch. Hydrobiol., *14:* 429-452. (Not seen.)

SMITH, F. G. W., R. H. WILLIAMS, and C. C. DAVIS
1950. An ecological survey of the subtropical inshore waters adjacent to Miami. Ecology, *31* (1): 119-146.

STIMPSON, MARGARET
1962. Reproduction of the polychaete *Glycera dibranchiata* at Solomons, Maryland. Biol. Stud. Cath. Univ. Am., No. 72. (Diss. Abstr., 23 [1]: 367.)

STOPFORD, S. C.
1951. An ecological survey of the Cheshire foreshore of the Dee estuary. J. Anim. Ecol., *20* (1): 103-122.

TARZWELL, C. M. and A. R. GAUFIN
1953. Some important biological effects of pollution often disregarded in stream surveys. (Reprinted from Purdue University Engineering Bulletin, Proceedings of the 8th Industrial Waste Conference, May 4-6, 1953.) U.S. Public Health Service, Environmental Health Center, Cincinnati, Ohio, 38 pp.

THOMAS, L. P., D. R. MOORE, and R. C. WORK
1961. Effects of Hurricane Donna on the turtle grass beds of Biscayne Bay, Florida. Bull. Mar. Sci. Gulf & Carib., *11* (2): 191-197.

THORP, E. M.
1935. Calcareous shallow-water marine deposits of Florida and the Bahamas. Pap. Tortugas Lab., *29:* 37-144.

THORSON, G.
1957. Bottom communities (sublittoral or shallow shelf). Mem. Geol. Soc. Amer., No. 67, Vol. 1: 461-534.

TURNER, C. H., E. E. EBERT, and R. R. GIVEN
1964. An ecological survey of the marine environment prior to installation of a submarine outfall. Calif. Fish Game, *50* (3): 176-188.

U.S. BOARD OF ENGINEERS FOR RIVERS & HARBORS
1922. Miami Harbor, Fla. Letter from the Secretary of War transmitting, with a letter from the Chief of Engineers, reports on preliminary examination and survey of Miami Harbor, Fla. Govt. Printing Office, Washington, D.C., 32 pp.

U.S. PRESIDENT'S SCIENCE ADVISORY COMMITTEE
1965. Restoring the quality of our environment. Report of Environmental Pollution Panel, U.S. Govt. Printing Office, Washington, D.C., 317 pp.

VOLK, R.
1907. Mitteillung über die biologische Elbe-Untersuchung des Naturhistorischen Museums in Hamburg. Verh. naturw. Ver. Hamb., *15:* 1-54. (Not seen.)

VOSS, G. L. and N. VOSS
1955. An ecological survey of Soldier Key, Biscayne Bay, Florida. Bull. Mar. Sci. Gulf & Carib., *5* (3): 203-229.

WALDICHUK, M. and E. L. BOUSFIELD
1962. Amphipods in low-oxygen marine waters adjacent to a sulphite pulp mill. J. Fish. Res. Bd. Can., *19* (6): 1163-1165.

WATSON, A. T.
1928. Observations on the habits and life-history of *Pectinaria (Lagis) Koreni.* Proc. Trans. Lpool. biol. Soc., *42:* 25-59. (Not seen.)

WEISS, C. M.
1948. The seasonal occurrence of sedentary marine organisms in Biscayne Bay, Florida. Ecology, *29* (2): 153-172.

WILHELMI, J.
1916. Übersicht über biologische Beurteilung des Wassers. Sber. Ges. naturf. Freunde Berl., *9:* 297-306. (Not seen.)

WILKINSON, L.
1964. Nitrogen transformations in a polluted estuary. Pp. 405-420 *in* E. A. Pearson (Ed.), Water pollution research, vol. 3. Pergamon Press, New York, vi + 437 pp.

WOODMANSEE, R. A.
1949. The zooplankton off Chicken Key in Biscayne Bay, Florida. Unpublished Master's thesis, College of Arts and Sciences, Univ. of Miami, Coral Gables, Fla., 110 pp.

WOODS HOLE OCEANOGRAPHIC INSTITUTION
1952. Marine fouling and its prevention. U.S. Naval Institute, Annapolis, Md., x + 388 pp.

WURTZ, C. B.
1955. Stream biota and stream pollution. Sewage ind. Wastes, *27* (11): 1270-1278.

YOUNG, P. H.
1964. Some effects of sewer effluent on marine life. Calif. Fish Game, *50* (1): 33-41.

Index